S0-BUA-863

SHAPING A NEW YOU

SHAPING A NEW YOU

The Truth About Weight Loss

James O. Hill, Ph.D.
John C. Peters, Ph.D.
Holly Wyatt, M.D.

Edited By
Jeffrey E. Modesitt

Published By
Shaping a New You Publishing Company
A Division of A. I. Solutions Inc.
Littleton, Colorado

First Edition 2003

Designed by Jeffrey E. Modesitt

Library of Congress Control Number: 2002116456

Modesitt, Jeffrey E., Editor
Shaping a New You: The Truth About Weight Loss—1st Ed.
Hill, James O.
Peters, John C.
Wyatt, Holly

ISBN 0-9726521-0-8

1. Diet 2. Weight Loss 3. Health 4. Title

This book is printed on acid-free paper.

10 9 8 7 6 5 4 3 2 1

ACKNOWLEDGMENTS FROM THE EDITOR

Consilience—a jumping together intellectually and geographically—describes the process required to produce this book. From the insight and support provided by Don Busby (Bakersfield), who knew that a common sense approach to weight loss and management should be offered as part of the *Shaping a New You* technology, to Rob Lowry, Massimo Torri, and Carmen Murray (Edmonton), all of whom worked diligently to develop the *Shaping a New You* software technology itself.

This book would not have been possible without the three authors and "lecturers"—Jim Hill (Centennial), John Peters (Cincinnati), and Holly Wyatt (Denver/Atlanta). They graciously found time in their unbelievable schedules to conceptualize and deliver the lectures from which the chapters in Part I and Part II were developed. They also served as the guiding lights in sifting truth from fiction in the morass of weight management information and misinformation that exists.

Without the support of Des O'Kell (St. Albert) and Angela Hatt (Edmonton), who shouldered much of my normal work load as well as their own, provided research assistance, and managed the book's production, this volume could not have been completed. To Brenda Belokrinicev (Edmonton) go my thanks for superb assistance in editing, both stylistically and contextually.

For my family and especially my wife, Leslie, who is my idea filter, stabilizer, reviewer, there are no adequate words to fully express my gratitude.

EDITOR'S NOTE

It is important that you seek—and heed the advice of—an expert, such as your physician and perhaps another qualified health/dietary/fitness professional, before you embark on any dietary or fitness program, and also that you have that expert follow your progress through the program. It is especially important for minors, pregnant women, breast feeding mothers, or anyone with any type of health condition or allergy, to seek professional medical advice before commencing any form of weight loss diet or program. Your physician or other qualified health/dietary/fitness professional can provide specific exercise and diet recommendations tailored to your personal or medical needs and circumstances.

The *Shaping a New You* book does not provide professional advice. All content, including any and all text, graphics, images, information, and other output (the "Content") is for general guidance purposes only. The Content is not a substitute for professional medical, dietary, or fitness advice, diagnosis, or treatment.

Links to Internet websites which may be provided in the book do not constitute endorsements of any information, products, or services which may be mentioned on such websites.

TABLE OF CONTENTS

ACKNOWLEDGEMENTS FROM THE EDITOR V
OVERVIEW 1
THE EXPERTS 13
INTRODUCTION 17
CHAPTER 1. CAN DIETING REALLY WORK? 19
CHAPTER 2. WHY ARE WE ALL GAINING WEIGHT? 35
CHAPTER 3. WHY WORRY ABOUT WEIGHT GAIN? 55
CHAPTER 4. THE BASICS OF ENERGY BALANCE 71
CHAPTER 5. LOSING WEIGHT 87
CHAPTER 6. WEIGHT MAINTENANCE 119
SANY SUPPORT SYSTEMS 139
PERSONALIZING YOUR WEIGHT LOSS PROGRAM 147
APPENDIX A. REFERENCE LIBRARY 171
COMPONENTS OF WEIGHT LOSS 174
CARBOHYDRATES 185
FATS AND FATTY ACIDS 190
PROTEIN AND AMINO ACIDS 204
VITAMINS 209
MINERALS 254
WEIGHT-RELATED ILLNESSES 279
SPECIAL TOPICS 290
APPENDIX B. WHAT IS COLORADO ON THE MOVE™? 301

OVERVIEW

Jeffrey E. Modesitt

Shaping a New You was written because the traditional approach to weight loss—the diet—doesn't work for most of us.

Shaping a New You is not the next new diet. A diet is what you eat. But changing what you eat is only one factor towards creating a lifestyle that will allow you to achieve and maintain a healthy weight.

This book is about a weight control and weight maintenance system that can work for everyone. It will give you the information you need to make the choices that will get you to the weight you want and keep you there.

Shaping a New You was written in conjunction with the development of the proprietary Shaping a New You (*SANY*) software that provides users with an unprecedented ability to plan and manage their weight loss and weight maintenance programs. There are two important steps in any weight management process: 1) understanding the complex factors that contribute to weight gain, and 2) managing the weight loss process in the easiest and most effective manner possible.

This book first explains why traditional approaches to weight loss are usually incomplete, often misleading, and almost always long-term failures. It then gives a detailed explanation of the elements required for a successful weight loss and management program, and explains why anyone who is serious about weight related issues should consider using the *SANY* software to increase their probability of success.

SANY'S UNIQUE CAPABILITIES

SANY software has unique capabilities and features that are based on the most current scientific and medical thought regarding weight loss and weight maintenance. These features include:

- true personalization of your weight management program based on **your** likes and dislikes;
- the ability to integrate the details of your diet and activities to produce personalized calculations of your Energy Balance;
- the ability to monitor your metabolic progress relative to your goals and make both dietary and activity recommendations to help you reach your goals;
- a "What if?" module that allows the user to preview dietary and activity choices to see their probable effect over a chosen period of time;
- a 3-D simulation module that produces an accurate visual representation of your progress or results of "What if" scenarios;
- nutritional feedback, including visual representation of your diet that alerts you to critical imbalances;
- the ability to sort and substitute recipes and individual food choices to maintain a balanced diet that meets your weight management goals or special dietary requirements;
- an interactive "Wizard" to guide you through all the software operations, and
- access to the *SANY* website, which is rich in helpful hints, new recipes, interviews with experts, links to other sites, and much more.

More information about this state-of-the-art software can be found at http://www.shapinganewyou.com.

Neither this book nor the *SANY* software mystifies weight loss by introducing yet another six-syllable super-hormone that magically explains why so many of us are overweight. Our goal is to de-mystify weight management and show you in simple terms how to achieve and maintain your ideal weight.

The Shaping a New You program was created by an international team. It integrates the knowledge and experience of world-recognized weight management experts with a proprietary artificial intelligence (AI) software system that can truly personalize your weight management program. The principles described in detail in this volume represent what we believe to be the soundest approach to weight loss available anywhere—the Intelligent Approach!

Achieving and maintaining weight loss is a holistic effort. Each of us needs to understand the critical relationship between our genetic makeup, our lifestyle—particularly food consumption and activity levels—and those physical and cultural elements that create the framework of our lives. If you undertake a diet without consciously dealing with all these factors, you dramatically diminish your chances of success.

This holistic effort may seem daunting, but it is worth it. Your goal of losing weight may be even more important than you realize!

THE WEIGHT AND HEALTH LINK

You may have noticed that public interest in weight-related issues has increased measurably over the last several years. The fact that we as a nation spend immense amounts of money to become thinner has focused the media spotlight on our overeating—but overeating probably isn't the real reason behind the current high level of coverage.

The driving force behind the coverage is the link between health and body weight. This is the truly important story!

The evidence is irrefutable. Increased weight is directly related to serious illnesses such as diabetes, heart disease, and cancer. The good news is that losing weight reduces the chances of developing these diseases—not a little, but a lot!

Losing weight is not the end of the story. Keeping your weight at your target level is as important as losing the weight, if not more so. You have to keep your weight off to maintain the important health benefits achieved through weight loss.

WHY A BOOK AND A SOFTWARE PROGRAM?

During the course of reading *Shaping a New You*, the reader will come to appreciate why achieving weight loss and keeping those lost pounds off are challenging tasks.

Why is it so hard to keep that weight off? What is missing? Simply this—most weight loss regimes lack supporting techniques.

In part, the culprit is our individuality. Until now, most diet calculations have been based on medical averages. Few of us fit those averages, so it is very difficult to know if a diet is likely to work until you've tried it. The *SANY* software harnesses an artificial intelligence capability that individualizes your weight management effort to your personal needs. In other words, information + artificial intelligence = success.

Information + Artificial Intelligence = Success

Wouldn't it be nice if you could know in advance what would happen if you changed your diet and activities? If you could design a diet and fitness plan based on your individual metabolism—*before* you commit to a course of action?

With the *SANY* software, you can! You can preview what you will look like six months, a year, or more in the future based on any number of "What if" scenarios. The *SANY* software allows you to simulate alternative events so that you can 'test' and choose the options that are best for you—so that you can choose the food and activities that are as close as possible to your current lifestyle.

The *SANY* software is the most advanced weight management program available today. It does not require memberships, monthly fees, or the sharing of your personal data with an unidentified someone over the Internet. While it is possible to lose weight by simply applying the principles discussed in the book, we believe that using the state-of-the-art *SANY* software will make your weight-loss process significantly easier to manage and will greatly increase your chance of success.

WHAT TO LOOK FOR

Shaping a New You contains a terrific amount of effective and easily understood information, but it is always helpful to have a framework with which to approach a book. This volume is organized around three guiding principles:

- The concept of **Energy Balance** underlies all weight loss and weight management programs.
- **Different weight loss** diets and **weight maintenance** programs work for different people.
- Our **unique individual metabolisms** and preferences defy the use of averages in creating effective individual weight management programs.

IT ALL STARTS WITH ENERGY BALANCE

Understanding how to balance the energy in the food you eat with the energy your body burns is the key to understanding weight management. Once you understand **Energy Balance**, you can choose a lifestyle that allows you to achieve the weight you desire.

Energy Balance is an easy concept to grasp. It involves comparing total energy intake (food and beverage consumption) with total energy output (your individual metabolism and physical activity pattern).

Sounds simple, doesn't it? It is! You will be able to harness this powerful concept quickly and effectively with the knowledge provided by the *Shaping a New You* software.

Energy In = Energy Out = Weight Maintenance

WHICH DIET IS RIGHT FOR YOU?

Bookstores and the shelves of public libraries are packed with volumes describing extraordinary diets. These diets run the gamut from simple to complex and from safe to risky. Some may reduce food costs, but many suggest the use of proprietary products that may double or triple your food expenses. Some are based on what are thought to be sound nutritional principles, while others suggest that new nutritional rules should be adopted.

This bewildering array of advice poses a dilemma for most of us. Where do you start if you are serious about weight control? How do you know if any of the fad diets are effective? If they are effective, are they safe? Which one is right for you?

DIETS WORK

The truth is that most diet programs on the market today do work. In case you think you might have misread that line, let us repeat it—most diets do work! Diets are designed to produce weight loss, and most do. As a matter of fact, we would be willing to bet that a large percentage of those who read this book have already lost weight using one or more of those diets.

Diets generally help us lose weight because most of us are willing to do exceptional things for short periods of time to achieve specific goals. We will eat grapefruit and hardboiled eggs, give up bread, pasta, and fruit, eat inordinate quantities of green beans, and so on ad infinitum. By doing these sometimes crazy and possibly harmful things, we are able to achieve gratifying short-term results.

But those gratifying results don't last. The majority of dieters are unable to maintain the required eating patterns and gradually—or not so gradually—they revert to the eating habits that are more closely aligned with their "normal" lifestyles.

Aligning your diet and your activities with your lifestyle is a critically important goal and one we will return to many times during the course of *Shaping a New You*.

Yes, diets work—repeatedly! They are so popular, in fact, that the expenditure for weight loss products and services now exceeds $100 billion yearly in the United States. That's about $350 per year for every resident of the United States—including babies.

However, if long-term weight loss **and** maintenance are the criteria for success, most of that hard-earned money is wasted. United States government figures show that 65% of Americans are overweight[1] and that many other nations either exceed this percentage or are in a race to catch up.

Americans spend $100 billion a year on weight loss products and services.

This is one of the great paradoxes of our culture. We spend huge amounts of money attempting to lose weight—only to regain it. Because self-esteem, health, and fitness are so important to us, we then search for another new and better diet and we start the process all over again.

Is there a rational explanation for this dilemma? More importantly, is there a real solution? We believe that effective weight control and maintenance is not only possible, but achievable if you understand and employ Energy Balance.

DON'T FORGET THE NON-DIETARY FACTORS

All of us will readily admit that we live in a very complex world. We are also increasingly aware of its interconnectivity. Benjamin Franklin recognized this when he wrote his famous lines illustrating a network of causation— "For the want of a nail the shoe was lost; for want of the shoe the horse was lost, for want of the horse the rider was lost!"

Diets are also "wanting" because they almost always fail to incorporate non-dietary factors that are critical to successful weight loss

---

[1] Flegal, K. M., Carroll, M. D., Ogden, C. L., & Johnson, C. L. *Prevalence and trends in obesity among US adults*. October 9, 2002. (reprinted) JAMA, 288, 1723-1727.

programs. Diets represent only one element of a successful weight loss and maintenance program.

ELEMENTS OF THE CONSPIRACY

The weight dilemma is the result of an unintended conspiracy.

No, we are not talking about some nefarious plot by the fast food industry. We're talking about a combination of diverse factors that influence weight control and management.

These factors fall into three general categories:

1) genetics,

2) personality and lifestyle, and

3) culture.

In the six lectures in Part I, our expert authors will first explain the elements that contribute to our weight epidemic and then will give you a powerful and effective program to help you achieve your desired weight.

This book is specially designed so that you can choose to read each lecture as a stand alone topic, or opt to read the whole book for a coherent approach to weight management. This format has meant that some content is included in more than one chapter, but it will allow you to use the book as a convenient reference during your weight management program.

Shaping a New You is a realistic program. Our experts will help you understand why you may **need** to pick a diet that fits your food preferences as closely as possible. You will learn that **what** you eat, more often than not, is not as important as **how much** you eat. This book will explain the relationship between diet and physical activity, and you will see that applying your new-found understanding of this relationship will give you the power to make choices that fit your lifestyle.

The authors acknowledge that there is much to be learned about weight loss and weight management. It is a complex subject. Fortunately, complete understanding is not necessary to achieve positive results.

We know that understanding and using Energy Balance, together with your consistent individual effort, are the keys to weight loss.

It is our goal to help you lose weight effectively.

The six chapters in Part I cover everything from whether diets really work to the real do's and don'ts of weight management.

Part II deals with personalizing your weight management program. We explain the *SANY* technology that will assist you in applying the authors' weight loss concepts more effectively. We provide a number of sample logs to assist you with your record-keeping, as well as an activity reference table that lists the energy consumption of numerous activities. You will also find helpful hints about reading food labels and estimating portion sizes.

If you would like to review a subject in more detail, Appendix A contains a comprehensive library on nutrition and related subjects.

Gain the power to make the right choices.

Part I

WEIGHT LOSS AND WEIGHT MANAGEMENT

THE EXPERTS

JAMES O. HILL, PH.D.

Dr. Hill is one of America's foremost experts in weight management. He has spent over twenty-five years understanding why people become overweight and how we can prevent and treat weight issues. He is well known as a co-founder of the National Weight Control Registry, which tracks over three thousand people who have lost weight and kept the weight off permanently. He was a member of the Expert Panel that developed the National Institutes of Health Guidelines for Management of Overweight and Obesity and Chair of the first World Health Organization Consultation on Obesity. He has served as President of the North American Association for the Study of Obesity and as Vice-President of the International Association for the Study of Obesity.

As Director of the Center for Human Nutrition at the University of Colorado, Dr. Hill has spent the past few years translating the science of weight management into simple programs that work in the community. He developed *The Colorado Weigh* and *The American Weigh,* formal behavioral weight management programs. Along with Dr. Peters, he developed *Colorado on the Move!* and *America on the Move!,* public health programs that use step counters to increase lifestyle physical activity in order to prevent weight gain.

Dr. Hill has published over two hundred scientific articles and lectured around the world on weight management. He has been featured in numerous television programs including a CNN special about obesity, *20/20,* and *Science Times* TV. He is a frequent source for stories on obesity written for the *New York Times, Washington Post, USA Today* and many magazines.

JOHN C. PETERS, PH.D.

Dr. Peters heads the Nutrition Science Institute of the Procter & Gamble Company and is an adjunct faculty member of the University of Colorado's Department of Medicine. He has conducted clinical research on food intake, Energy Balance, and obesity for over twenty-five years, has lectured internationally, and has published over ninety scientific articles.

Dr. Peters sits on numerous scientific advisory boards including the Arkansas Children's Hospital Research Foundation, the Cincinnati Children's Hospital weight management program, and the University of Colorado Center for Human Nutrition. He is active in local and national health promotion, and plays a leading role in several non-profit organizations dedicated to health promotion and disease prevention. He serves on the Executive Committee of the Partnership to Promote Healthy Eating and Active Living, is Chairman of the Friends of the University of Colorado Center for Human Nutrition, and is Vice Chair and former Executive Director of the Physical Activity and Nutrition Program of the International Life Sciences Institute Center for Health Promotion. Dr. Peters holds B.S. and Ph.D. degrees in biochemistry from the University of California at Davis and the University of Wisconsin–Madison respectively.

HOLLY WYATT, M.D.

Dr. Wyatt is an Assistant Professor in the Department of Medicine, Division of Endocrinology, Metabolism and Diabetes at the University of Colorado Health Sciences Center. She is currently a physician and clinical researcher at The Center for Human Nutrition in Denver, Colorado.

She received her B.A. from the University of Texas and her M.D. from the Baylor College of Medicine in Houston, Texas. She completed both an Internal Medicine Residency and an Endocrinology and Metabolism Fellowship at the University of Colorado Health Sciences Center. She is board certified in Internal Medicine.

Dr. Wyatt's research interests are in understanding energy and macronutrient balance in obese individuals and individuals who successfully maintain a weight loss (reduced-obese). She has expertise in the science of measuring energy expenditure in human subjects. She received both a National Service Research Award and a Patient-Oriented Research Career Development Award from the National Institutes of Health (NIH) to study energy metabolism and metabolic factors associated with weight loss in reduced-obese subjects.

Dr. Wyatt is the National Coordinator for the Centers for Obesity Research and Education (CORE). CORE is a national organization mandated to provide educational workshops to health care providers on obesity assessment and management in their practice. She is also Medical Director of *The Colorado Weigh* weight management program. Dr. Wyatt has helped hundreds of overweight and obese people lose weight and keep it off.

INTRODUCTION

The following six chapters were adapted from a series of lectures given by James Hill, John Peters, and Holly Wyatt. As you have read, their qualifications are exemplary. Our authors are not theory-bound scientists and doctors who never get out of the laboratory. Their combined experience represents more than five decades of helping real people lose weight and maintain their weight loss.

The lecture series is designed to lead the reader through the weight loss and management process. If you are looking for a "diet" book that dictates your eating patterns, this book is not for you. Effective weight loss and weight management require more than a diet plan, particularly if that plan is one that a dieter can stomach only briefly (pun intended) before reverting to their previous habits of overeating.

In the Overview, we suggested some of the reasons diets work only in the short-term but fail in the long-term. Our authors will now guide you through the critical elements of weight loss and provide you with proven techniques that will help you modify your current lifestyle and achieve your weight objectives.

This book does not contain recipes, a list of "must do" exercises, or any "guaranteed instant weight loss" gimmicks. *Shaping a New You* is about reality—your reality. As you read these chapters, you will come to appreciate that you can make choices that are compatible with your lifestyle. You will be able to pick a diet that fits your preferences and choose activities with which you are comfortable.

You may have questions regarding nutritional basics. Appendix A contains an extensive "Reference Library" designed to help you answer questions regarding medical terminology and nutrition. The Library also contains lists of further references should you wish to learn more.

If the concepts in this book make sense to you, you owe it to yourself to maximize their application through the use of the revolutionary Shaping a New You software. Find out more in Part II.

CHAPTER 1. CAN DIETING REALLY WORK?

John C. Peters, Ph.D.

If you bought this book, you are probably interested in weight management and considering beginning a diet. Many of you may already have been on a diet at some time in your life. It may be that whatever you did in the past did not work as well as you would have liked and you might be asking yourself the question, "Can dieting really work?"

When we examined the dieting landscape in this country, we found a good news/bad news state of affairs. Let me tell you what we found.

ALMOST EVERYONE IS GAINING WEIGHT

If we look not just at individuals, but at our entire society, we find that almost everyone is gaining weight! Weight gain is occurring in people whether or not they are already overweight. It is occurring in people of all ages. It is occurring both in cities and in rural areas. It is occurring in all ethnic groups. And it is occurring in our children!

About 65% of the adult population in the United States—that's about 122 million people!—are overweight or obese. And this number is increasing every year.

People are considered overweight if their body mass index (or BMI) is greater than 25 kg/m^2, or obese if their BMI is greater than 30. Don't worry about trying to calculate the kg/m^2 formula. We'll talk more about it in Chapter 3 and we'll provide you with an easy-to-use chart to determine your own BMI (see page 57).

Have a look at the table below.

Change in overweight and obesity in the U.S.[2]

Overweight (People with a BMI > 25)	Obese (People with a BMI > 30)[3]
65% of adults age 20 to 74 About 122 million	30.5% of adults age 20 to 74 About 59 million
61.9% of females 60 million	33.4% of females 32.5 million
67.2% of males 62 million	27.5% of males 26.5 million
14% of all children & adolescents	

Let's focus on the implications of these numbers. About 60 million women and 62 million men in this country have a weight problem. About 59 million American adults (30.5% of the country's population) are obese. These people all have a body mass index greater than 30 (you'll be hearing a lot about this particular number).

Even more frightening is the fact that we are seeing an increasing number of overweight and obese children. We estimate that about 14% of children and adolescents in the United States are either overweight or obese. This number is also increasing every year. The situation could not be more alarming—two thirds of our country is overweight or obese and only about one third is able to maintain a healthy body weight.

What could be causing this disturbing state of affairs?

---

[2] Flegal

[3] The symbol ">" means "greater than." The symbol "<" indicates "less than."

How did most of us start carrying around so much body weight? We can get some clues as to how this happened by following the changes in body weight of people who participated in long-term research studies. One good example is the Coronary Artery Risk Development in Young Adults (CARDIA) study. This study followed young men and women for fifteen years to learn about the development of heart disease. Although the study was not designed specifically to study weight, participants were weighed and measured periodically.

Recently, CARDIA investigators examined weight gain in their subjects. They found that all of the groups—Caucasian men, Caucasian women, African-American men, and African-American women—had gained weight during the period of the study. African-American women had gained the most weight, at over two pounds a year, while Caucasian women had gained the least, at just over one pound each year.

The important message from this study is that over 75% of these typical American men and women gained between one and three pounds every year for fifteen years! And that made them overweight or obese.

On average, Americans gain 1-3 pounds every year!

These results paint a dismal picture. They show that almost everybody is at risk for weight gain.

You're not alone in your quest to learn how best to manage your body weight, and we're going to talk more about how you can achieve success.

UNDERSTANDING ENERGY BALANCE

The American obesity epidemic that we have just described seems to have come from a weight gain of only one to three pounds per year.

How can this be? This is where we learn more about the concept of **Energy Balance**. Your weight is the result of your Energy Balance over time.

The average weight gain we described above—one to three pounds—was the result of a very small positive Energy Balance. It took only an extra 10 to 30 calories of positive Energy Balance a day to put on that weight. Yet over 25 years, that little bit of extra food eaten each day has made Americans 25 to 75 pounds heavier!

Most weight gain is caused by a small degree of positive Energy Balance.

Let's explain Energy Balance. The food you eat contains energy, and your body burns this energy to keep you functioning and to provide fuel for your physical activity. Some energy can be stored in your body—most of this is stored in the form of body fat. Each pound of body fat stored in your body represents about 3,500 calories.

Our bodies store fat to ensure we will have a supply of energy when we need it. The problem is that too many people are storing too much energy as body fat.

This body fat is accumulating relatively slowly, at the unnoticeable—at first!—rate of one to three pounds each year. At this rate most of us don't pay attention to the weight gain. Then, one morning we wake up and see we are fifteen, twenty, or thirty pounds overweight.

We are overweight simply because our lifestyle has produced (and is probably continuing to produce) a positive Energy Balance. Lifestyle, as used in this book, is not just our food consumption, but also includes physical activity, cultural, and environmental factors. This is a critical point to understand.

Energy In = Energy Out = Weight Maintenance

Your Energy Balance is the difference between the total amount of energy you take in by eating food during the day (energy intake) and the total amount of energy your body burns during the day (energy expenditure). If your total energy intake is exactly equal to your total energy expenditure, your body weight (energy stored in your body) will not change. This situation is called zero Energy Balance.

The only way you can gain weight is for your energy intake to be higher than your energy expenditure (creating a positive Energy Balance). The only way you can lose weight is for your energy expenditure to be higher than your energy intake (negative Energy Balance).

Energy In > Energy Out = Weight Gain

Energy Out > Energy In = Weight Loss

For you to reach and maintain the weight you want, you must understand how to take control of your Energy Balance.

TEN CALORIES A DAY

Now back to our weight gain problem. It only takes about ten extra calories each day (in other words, a positive Energy Balance of only ten calories) to gain one pound in a year.

You can do this by eating just one extra lifesaver a day, or by walking 250 steps (about a tenth of a mile) less than usual each day.

The important point is that **most weight gain is caused by a small degree of positive Energy Balance**.

Looking at the problem in this way, we begin to see that weight management is often a matter of simple choices. If we want to maintain our current weight, we need to manage our food intake so that it balances our energy expenditure. Weight loss requires either a reduction in energy intake, an increase in energy expenditure, or a combination of both. Once you understand Energy Balance, you have the power to take control of your body weight.

In the beginning of this chapter, we asked if dieting can really work. We told you the numbers—65% of Americans are overweight or obese and the situation is getting worse.

At the same time, more than half of the women in the country are on a "diet." These women are trying hard to change their body weight.

Despite all this effort, as a nation we continue to gain weight every year. While many people are trying to lose their excess weight, few are successful in losing that weight and keeping it off. In fact, the numbers show that 80% to 95% of the people who have successfully lost weight gain it all back!

---

1 Extra Lifesaver Each Day Over Energy Balance

= 1 Pound Gained Each Year

We are stuck in an endless cycle of lose it—gain it—lose it—gain it. Some people have said they've lost a "ton" of body weight in their life, and in some cases that may be true. But they haven't been able to keep it off.

At this point, you might be thinking that the whole thing sounds hopeless. Our intent is not to discourage you.

The answer to the question, "Can dieting actually work?" is a conditional yes—if you understand Energy Balance. If you do not understand Energy Balance, you are probably doomed to repeat the endless weight loss/weight gain cycle.

Managing your weight is not just about dieting, it is about losing weight by establishing a negative Energy Balance and then adopting a lifestyle that allows you to keep the weight off. It does not have to mean changing your whole life. It simply means making small changes and allowing those changes to become a permanent part of your routine. In most cases, small changes work better than large ones—because major lifestyle changes are incredibly hard to maintain.

THE NATIONAL WEIGHT CONTROL REGISTRY

Nearly everyone knows someone who's succeeded at losing weight and keeping it off. While few of these people make it into the scientific literature, they do exist.

In the last few years, scientists have put together a database of people who have successfully lost weight and kept it off. The National Weight Control Registry (NWCR)[4], which you'll be hearing much more about later in the book, is a list of over 3,000 people who have successfully lost at least 30 pounds and kept it off for at least a year. People can register only if they have successfully lost weight and kept it off.

The average weight loss in this group is 67 pounds and the average time that they have kept the weight off is 6 years.

This large, diverse group of people has truly succeeded at adopting behaviors that allow them to maintain the weight they want over the long-term. These individuals used many different strategies to lose weight and keep it off.

You can lose weight and keep it off.

When we analyzed the data compiled by the NWCR, here's what we found. Although registrants had successfully lost weight through a variety of formal and informal weight loss programs, these diverse programs had certain general behaviors in common:

- Participants weighed themselves often.
- Participants watched their fat and calorie intake.
- Participants ate a sensible breakfast each day.
- Participants were physically active.
- Participants kept a journal.

---

[4] National Weight Control Registry. Available at: http://www.uchsc.edu/nutrition/nwcr.htm

The point is—no one size fits all. Weight loss is very personal, and you have to find the program that works for **you**.

It is possible to succeed at dieting once you figure out the best plan for you. The intent of this book is to help you do exactly that.

THE ONE PRINCIPLE TO FOLLOW FOR WEIGHT LOSS

How do you find the right way to lose weight and keep it off?

To answer this, we have to get back to Energy Balance. Any time you create negative Energy Balance, you lose weight. The greater the negative Energy Balance, the greater the weight loss. In fact, **creating negative Energy Balance is the only non-surgical way (liposuction is an example of a surgical method) to lose weight**.

You cannot lose or gain weight if your energy intake and energy expenditure are equal. You are at zero Energy Balance.

Energy In = Energy Out = No Weight Loss or Gain

Energy In < Energy Out = Weight Loss

Energy Out > Energy In = Weight Loss

You cannot lose weight if your energy intake is greater than your energy expenditure. You have a positive Energy Balance and will gain weight.

The only way to lose weight is to burn more calories than you eat. That is, you lose weight only when you have a negative Energy Balance.

It's quite simple. It works the same in everybody. No two people may be alike, but this principle works the same way in everybody.

CREATING NEGATIVE ENERGY BALANCE

One pound of body weight represents about 3,500 calories of energy. If you want to lose a pound, you have to burn 3,500 more calories than you eat over some period of time. You can choose how long that period of time is. For example, to lose one pound in a week you have to eat 500 calories less than you burn, or burn 500 calories more than you eat each day.

It doesn't matter how you do it. Your goal is to create that 500 calories-a-day negative Energy Balance. To do this, you can combine different proportions of eating less and moving more. We'll talk more about the impact that each of those changes to your lifestyle can have.

Remember—one pound of body weight is about 3,500 calories.

- To lose a pound, you need to burn 3,500 calories more than you take into your body.
- You can do this by eating less, by becoming more active, or through a combination of the two.
- To lose one pound per week, you would need to burn about 500 calories more per day than you take into your body.

CHOOSING THE RIGHT APPROACH

There are many different ways to create negative Energy Balance. The choices you make are important.

Negative Energy Balance can be created by an infinite number of combinations of eating less and moving more. Your challenge is finding the right plan for creating negative Energy Balance—a plan that fits your individual physiology and your unique lifestyle.

What is the mix of eating less and moving more that's compatible with your life? There is no magic pill, there is no single food, no single exercise that will help you to lose weight and keep it off forever. You have to find the balance of eating less and moving more that works for you.

WHERE MOST DIETERS FAIL

After you have lost the weight you wanted to lose, you no longer need to maintain a negative Energy Balance. To maintain a steady weight, you need to be in zero Energy Balance.

Here is where most dieters fail. People believe that once they have lost the weight they can return to their "normal" lifestyle, adopting the same behaviors they had before they lost weight.

In fact, many dieters seem to believe that by losing weight they have somehow been "cured" and that their metabolism has been increased. Post-dieters tend to act as though their successful "cure" has changed them, as if they can now behave like the thin people they know—who don't ever seem to have to worry about what they eat.

In reality, their metabolism is **lower** after their weight loss than it was when they were heavier. They are smaller now, and their bodies know it and have adjusted their metabolisms accordingly.

These successful dieters will need to make permanent adjustments in their behavior to maintain their reduced body weight. If they return to their pre-weight-loss lifestyle, it should not be surprising that they will return to the greater weight that was associated with that lifestyle.

After successful weight loss, then, your challenge is to keep the weight off. This requires the state of Energy Balance that we refer to as **zero Energy Balance**. Now your goal is to make sure that your energy intake is the same as your energy expenditure.

Energy In = Energy Out = Weight Maintenance

In later chapters, we will show you how to think about managing your body weight in much the same way you manage your finances.

To achieve the bank account you desire, you have to manage your income and expenditure. Similarly, to achieve the weight you want, you have to manage your energy intake and energy expenditure. While the goals are opposite (you want to increase your money and decrease your weight), the principles needed to reach the goals are exactly the same.

Once you understand this, you can develop a simple plan to accomplish your goals.

BECOMING AN EXPERT

This book can help you become an expert in food and physical activity. You need to learn about where the calories are if you're going to change the input side of the Energy Balance equation. You must also understand that food intake is only one side of the Energy Balance equation. It is equally important to recognize how to modify the energy expenditure side of the Energy Balance equation.

Your body burns calories to perform work or physical activity. Increasing physical activity is almost certainly a strategy you will want to consider as you manage your Energy Balance and your body weight. Your body burns some energy even when you are sitting around doing nothing.

You will learn more about this later in the book when we talk about your resting metabolic rate or RMR. Part of our intent in writing this book is to increase your knowledge about the relationship between food and physical activity.

The changes in Energy Balance you can produce by changing your food intake are different from the changes you can achieve by manipulating your physical activity. For this reason, certain strategies may be better for weight loss than for keeping the weight off.

A NINE HOUR BIG MAC MEAL

To illustrate this, let's consider the energy in a Big Mac. A seventeen-year-old boy can probably eat an entire Big Mac in about three minutes. A Big Mac has about 560 calories in it, so this young man would be taking in an average of 187 calories a minute. That is a lot of energy entering the body during a brief period of time.

Energy intake versus energy expenditure: No contest

Activity	Calories gained or lost	Duration	Efficiency (kcal/min)
Eat a Big Mac	Add 560 calories	3 minutes	Add 187 calories per minute
Exercise required to achieve Energy Balance at 70% of maximum effort	Burn 560 calories	60 minutes	Burn about 9 calories per minute
Exercise required to achieve Energy Balance at 40% of maximum effort	Burn 560 calories	90 minutes	Burn about 6 calories per minute

Let's consider how long it would take for your body to burn this amount of energy. If you are just sitting around, it will take you about three hours to burn off the Big Mac. But let's say you want to burn it off through exercise. If you exercise pretty vigorously, say at 70% of your maximum capacity, it will take an hour for you to burn the energy in the Big Mac. Even during strenuous exercise, you are only burning about 9 calories a minute.

If our seventeen-year-old "super-sizes" the meal with a non-diet drink and fries, the calorie count rockets to about 1,825. It would take over nine hours of relatively sedentary activity, or about three hours of exercise, to balance the energy account.

Most of us aren't able to exercise at 70% of our maximum capacity for even a full hour. This means that a more realistic effort to balance our intake energy from just the Big Mac—not the entire meal—through exercise would be closer to two hours.

This fast-food favorite illustrates why it is very difficult to lose weight through exercise alone! Losing one pound of weight—fat mass, not water—would require running about 35 miles. A marathon is only 26 miles.

---

1 Pound = 1.34 Marathons

A better choice might be to eat a cheeseburger instead of the Big Mac. The cheeseburger contains only about half as many calories, so it would require about half the time for your body to burn the calories.

Once you understand energy and Energy Balance, you will be better able to make simple choices that will have major effects on your body weight. We believe that, when you understand Energy Balance, you will be empowered to make intelligent choices about your lifestyle and you will know how to plan a strategy for managing your weight. There will be no need for you to resort to fad diets or weight loss supplements.

You can achieve and maintain the weight you want by making decisions such as choosing to eat a cheeseburger instead of a Big Mac. You will be surprised at how making small changes in your current lifestyle will allow you to lose weight and keep it off.

Remember, the average weight gain over a one year period is the result of only about 10 to 30 calories of positive Energy Balance each day! Once you know about Energy Balance, you can manage your weight in a sensible and achievable way by taking conscious control of the primary factors contributing to your unique Energy Balance.

BE AWARE OF YOUR ENVIRONMENT

Many people gain weight because they fail to consciously examine environmental factors and societal values that have an impact on their activity levels and eating habits. These environmental elements will be described in more detail in the next chapter, but, generally, they can be categorized as factors that encourage a combination of eating too much food and being physically inactive.

The people who maintain a healthy weight are often the ones who make a conscious effort to counter the influence of their environment. Choosing the cheeseburger over the Big Mac is an example of an intelligent choice that is probably easy to accomplish without a measurable reduction in satisfaction.

By choosing the cheeseburger over the Big Mac, you save 280 calories and up to 45 minutes of physical activity. A simple choice—yet too many people choose the Big Mac.

To successfully change your Energy Balance, you will need to become an expert at managing the environmental pressures that make you want to eat the Big Macs of the world. You will also need to learn to manage the environmental factors that make you want to sit rather than to move.

Your brain is the most important tool you have in your battle with your weight. By making informed choices, you will be able to make small changes in your lifestyle that will bring you positive results.

TRADITIONAL DIETING IS A "TIME OUT"

Managing your weight involves psychology as well as physiology. Let's consider your previous weight loss attempts. You may have been on ten different diets in the past and lost weight on every one of them. If your measure of success was temporary weight loss, they may all have worked. However, if keeping the weight off was your measure of success, then none of them worked.

Why did these diet plans fail? What did you do wrong?

You chose to try to do the nearly impossible.

It is easy to make big changes in your life temporarily. You can follow the grapefruit diet, or even a beer diet, for a period of time—and you will probably lose weight.

But it is unlikely you will be able to stay with such diet plans very long. They work in the short-term because they represent such a big change from the way you eat now.

They fail in the long-term for exactly the same reason. We can make temporary major changes in our lives, but it is much harder to make permanent major changes.

Traditional dieting is like taking a time-out from your regular life. You set your regular life aside and say, "Okay for the next three months, I am going to go full speed ahead on whatever the program is."

Disruptive changes in eating habits generally do lead to some amount of weight loss. But when you reach your weight loss goals, the problem becomes—what to do next?

No one can maintain a grapefruit or beer diet forever. Most people end up going back to their regular lives—but their regular lives were what caused them to gain weight in the first place. It's no surprise that most dieters regain their lost weight.

INTELLIGENT CHANGES ARE OFTEN SMALL CHANGES

The secret to weight management success is not making huge lifestyle changes, because they're difficult to sustain. Your goal is to find the right blend of small changes—changes you can live with, not changes that are just temporary time-outs.

Strategies such as artificially structuring your eating, totally avoiding temptation, and limiting certain foods all work very well—in the short-term. It is difficult to maintain these strategies for long periods of time, though, and usually the weight is regained when the person can no longer live up to these strategies.

A QUICK REVIEW

Long-term success in weight management is possible. Success involves both losing weight and keeping it off. The key to weight loss is creating negative Energy Balance and the key to long-term success is maintaining zero Energy Balance by adopting behaviors that you can maintain forever. Making small changes in your behavior patterns that are consistent with your lifestyle preferences will usually work far better than making massive changes that are much more difficult to sustain.

What is the right blend of eating less and moving more for you—the one that you can keep up for the rest of your life?

Think of having two different strategies for controlling your weight. One strategy is eating less and the other is moving more. You can reach your goals by using either strategy alone. For example, if you do not want to move more, you could manipulate your Energy Balance by only changing your diet. Alternatively, if you do not want to change your food intake, you could do it by only increasing your physical activity.

When we identify people who have been successful in weight management, however, we find only a few who have done it using a single strategy. Almost all of those who experienced success at both weight loss and weight maintenance modified both diet and physical activity.

Less Energy In + More Energy Out = Negative Energy Balance

You can make your own decision on the right proportion of change in each component, based on your likes and dislikes. Maybe it's moving 20% more and eating less. Maybe it's half and half—a little more activity, a little bit less food. The key is getting in touch with yourself and with your lifestyle, and figuring out the right blend of the two strategies—eating less and moving more—that is going to create a negative Energy Balance for you. This combination of behavioral changes must be one you can maintain for the long-term.

You also have to decide how much weight you want to lose and how quickly you want to lose it. Each decision you make has consequences for your Energy Balance and for your body weight. The tools in this book take you through these decisions and discuss the consequences of various choices. Applying these tools will help you to make informed choices that will allow you to manage your weight more comfortably.

CHAPTER 2. WHY ARE WE ALL GAINING WEIGHT?

John C. Peters, Ph.D.

Before you start your weight loss program, it may be helpful to discuss critical biological and societal factors that might have contributed to your weight gain. If you understand these sometimes subtle—but powerful—influences, your future weight management efforts will have a much better chance of success.

The factors we will discuss in this chapter involve the relationship between your genetic background and your current physical and cultural environment. Their combined influence crosses virtually all societal and cultural boundaries in the United States and has resulted in a virtual epidemic of overweight and obesity.

In October 2002, the United States government released its 1999-2000 figures showing that 65% of the adult population is either overweight or obese. This represents an increase in the proportion of the population who weigh more than is healthy of 4% in only two years. In fact, the proportion of the population considered overweight or obese has been steadily increasing for many years. Why is this happening—here and in many other countries around the world?

If you are asking yourself whether we are experiencing some sort of epidemic of bad willpower, you are not alone. Everyone is concerned. But that's not the case. We do have an epidemic of obesity, but it is not due simply to people having inherently low willpower.

We are trying to lose weight as no other nation in the history of the world has ever before tried. With over 50% of women in the United States on a diet as you read this and a national expenditure for weight management of over $100 billion yearly, it is obvious that there is more to the problem than willpower alone. In this chapter you will see how our willpower is actually being negated by some very powerful biological and societal influences. You must refuse to be the victim of these influences. Use your knowledge of Energy Balance and take control.

IDENTIFYING THE NEGATIVE INFLUENCES

The Center for Disease Control and Prevention (CDC) collects state-by-state information about obesity. Each year, the CDC calls a representative sample of people from each state and asks questions about their health and their lifestyles.

Body mass indices, or BMIs, are calculated from the information collected on height and weight. BMIs allow the CDC to track obesity over the years.

Keep in mind that this information likely underestimates the real prevalence of obesity, as people tend to under-report their weight and over-report their height. Nonetheless, because the information is collected in the same way each year, the relative changes in obesity are likely to be accurate.

The figures are startling. Have a look at the maps on the following page. The map on the left shows the prevalence of obesity in 1991. Recall that obesity means a body mass index of greater than 30. You can see that in 1991 only four states plus Washington D.C. had an obesity prevalence between 15% and 19% and no state had an obesity prevalence greater than 20%.

The map on the right shows the situation only nine years later, in 2000. In just nine years, the situation deteriorated dramatically. Only one state, Colorado, had a prevalence of obesity less than 15% and all other states had obesity prevalence above 15%.

Obesity Trends Among U.S. Adults

1991 **2000**

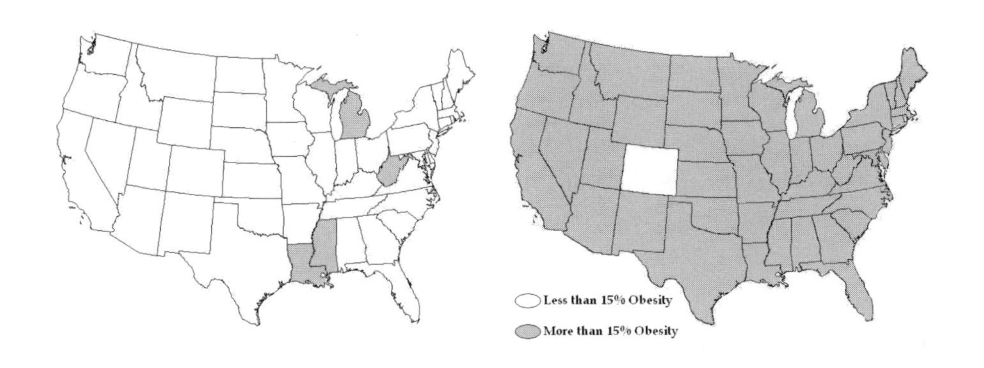

Source: Centers for Disease Control (CDC)

It seems that almost no one is immune to weight gain.

What is causing this epidemic of weight gain? The most likely explanation for the dramatic increase in obesity is that something in the environment is affecting everyone, no matter who they are or where they live. Whatever it is, it's making it very difficult for most people to avoid gaining weight over time.

It is clear that everyone must take personal responsibility for their own lifestyle and their own body weight. However there is a lot more involved here than personal responsibility.

The society and cultural setting we live in has a strong influence over our behaviors. Often this influence goes unnoticed.

We are also dealt a genetic code that has been perfected over millions of years. Our genetic code is the result of living in and adapting to our environment. But that environment, the one that has shaped our evolution over the centuries, has essentially disappeared in the last fifty years.

If we are to be successful in our weight management efforts, we need to understand the influence of these two factors on our lives.

EVOLUTION AND BIOLOGY

Why is obesity such a problem today when it has never before in our history been a serious problem? Is it something about our physiology that's changing? Is it the environment around us? Is it the way we live? Or is something about our inherent values as a culture contributing to this problem?

You might wonder if becoming obese is a natural part of our evolution. Should we just accept that obesity is a stage in the evolution of mankind? It's an important question and one we'll attempt to answer in the next section.

THE BIOLOGICAL FACTOR

Let's first look at our biology. The genes with which you were born clearly affect your physiology and influence what you look like. Chances are that you resemble one or both of your parents in your height and weight. Our genes contribute importantly to explaining why people vary so much in their heights and weights.

Your genes affect how your body works—how it processes food, and how it burns nutrients for energy. Because everyone has a different genetic pattern, everyone looks slightly different and everyone's body functions slightly differently.

It is impossible, however, for alteration of our genes to explain the weight gain we have experienced as a population over the past two decades. The genetic pattern of a population changes very slowly, over many, many thousands of years. Genetic changes begin with an individual and then spread gradually through multiple generations. No change in the genes of our population can explain the epidemic of obesity that has arisen over a few decades.

If you think about your friends, they probably vary widely in body weight. Genes can contribute to these differences. However, most of your friends probably weigh more today than they did ten years ago—that is not because their genetic structures have changed.

It is because of the combination of old genes and a new environment. The fact is, our environment has changed considerably over the past two decades, and the changes are pushing us all to gain weight.

While we cannot attribute the weight gain to a change in our genes, it is true that our genes are a major factor in the national weight epidemic. Millions of years of evolution have made the human body very efficient at storing fat. Until the last few decades, our environment demanded this ability as a critical survival skill. Even three decades ago, the combination of high physical activity levels and food intake were fairly balanced.

That balance has disappeared as our environment incorporates more and more laborsaving technology.

ISOLATING THE EFFECTS OF CULTURE

We can use a real-life example to demonstrate how genes and the environment interact to affect whether or not obesity develops. The Pima Indians, a group of Native Americans, live on a reservation in southern Arizona. About 400 years ago, their ancestors migrated there from the northern mountains of Mexico. A small community of Pima Indians also remained in northern Mexico. Let's have a look at how the experiences of the two groups compare.

These two groups have virtually the same genetic make-up but have very different lifestyles. The Pimas who live in northern Mexico are subsistence farmers. They have a very active lifestyle and, by necessity, they are largely vegetarian. Their diet is one that most nutritionists would consider to be very healthy.

Environment matters.

The Pimas in southern Arizona, on the other hand, live on a reservation, are very inactive physically, and eat a typical North American diet, high in fat and calories. Most nutritionists would consider their diet to be relatively unhealthy.

These two groups, then, have similar genetic codes, but very different lifestyles. Interestingly, their body mass indices are also very different. Most of the southern Arizona Pimas are obese and suffer from weight-related diseases. The northern Mexico Pimas, on the other hand, enjoy dramatically lower levels of obesity and are presumed to have lower levels of obesity-related diseases. However, the BMI of the northern Mexico Pimas was higher than might be expected from a population eating a low calorie density diet and engaging in high levels of physical activity.

The Pima Experience: Environment Makes a Difference

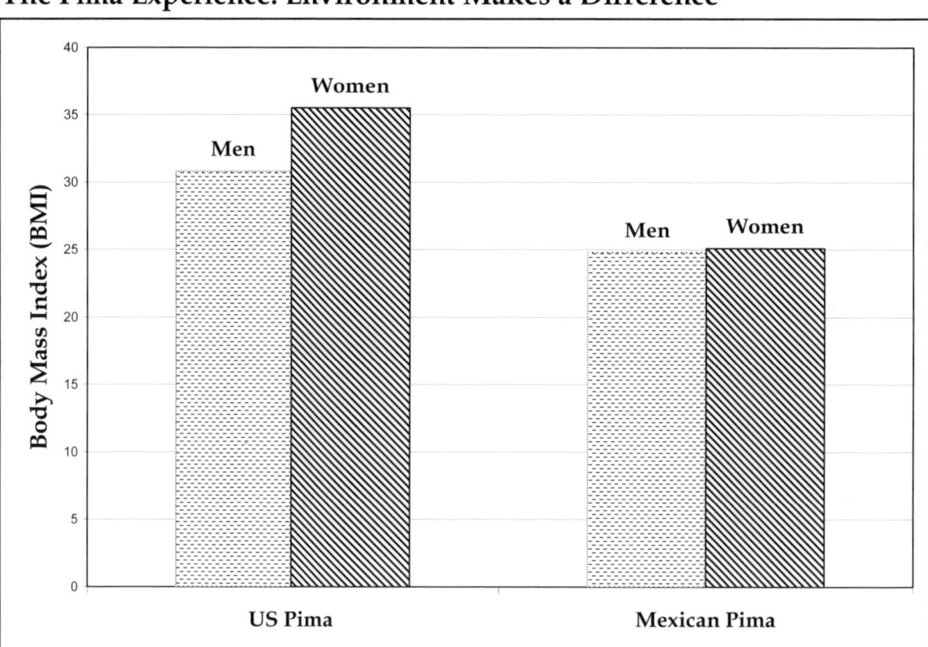

Adapted from Ravussin E., Valencia, M. E., Esparza, J., Bennett, P. H., & Schulz, L. O. Effects of a traditional lifestyle on obesity in Pima Indians. *Diabetes Care.* 1994;17:1067-1074.

This example illustrates two points. First, it illustrates the power of the environment to affect body weight and obesity, since the genetic make-up did not differ significantly between the two groups of Pimas. Second, the higher BMI of the northern Mexico Pimas, relative to physically active people on similar diets, shows that genes also influence body weight.

A MISMATCH OF GENES AND ENVIRONMENT

Given the substantial difference in BMIs between the two Pima groups, we must accept that changes in our environment, given a stable gene pool, encourage weight gain and obesity.

In fact, we Americans seem to have a mismatch between our genes and our environment. The human body has evolved over millions of years. Its design equips it to deal with situations that don't exist in this country today in any major way.

The human body was designed to deal with a scarcity of food, and with a harsh environment that demanded high levels of physical activity—hard labor—just to get through the day. Our ancestors had to find and prepare food, find and prepare shelter, and protect themselves and their families from the environment. All of this physical activity burned a tremendous amount of energy. Our ancestors needed very high levels of food intake just to maintain enough body weight to survive and support reproduction.

Genes matter too.

Over millions of years, the body developed redundant mechanisms to ensure that we got enough food to survive, given our high levels of energy expenditure. It made sense for us to eat whenever food was available. If there was more food available than we actually needed, we ate more anyway—because we never knew when the next meal would come. We stored extra food as fat, to be used as an energy source during periods of food scarcity. The notion of "clean your plate"—eat whatever you can whenever you can—may have been built into our behavior. It seems to have become instinctual.

Similarly, because we had to be so physically active to survive, we may not have developed systems to encourage us to be physically active when we did not need to be. Rather, it was more likely that we developed a tendency to rest and conserve energy whenever that was possible. Again, this may have become instinctual.

Are our instincts killing us?

Such instincts would have served us well in earlier environments. In fact, the population today probably reflects those with genes that were particularly strong for eating (and storing energy as fat) and resting (insuring conservation of energy). People with different genetic patterns likely did not fare so well during times of food scarcity. It is a simple matter of survival of the fittest—that is, survival of those best equipped for life in their environment.

INSTINCTS LEAD US ASTRAY

Our instincts for eating and resting do not serve us well in the environment that we have created today in the United States. This environment is very different from any other we have ever experienced.

In today's environment, we are surrounded by ready-to-eat foods and we require almost no physical activity to get through the day. If we rely on our instincts, we will eat all of the time and spend as much of the day as possible resting.

We are not trying to gain weight. Weight gain is occurring as an unanticipated consequence of the fact that we, as a society, have achieved our dreams of having food always available and of not having to engage in high levels of physical activity for survival. You could say we are victims of our own success.

Eating all you can is a survival technique!

Great-tasting food is everywhere. It's inexpensive, and it comes in gigantic portions. You cannot get away from food whether you are at the airport, the shopping mall, the gas station, your desk, or the bookstore. Everywhere you go, you can get something to eat.

Food consumption represents only half of the instinctual betrayal, however. Most of us work at jobs that require very little physical activity. We have more free time than ever, but we spend most of that time sitting—watching television, playing video games, surfing the Internet. We live in an environment that encourages energy intake and discourages physical activity. It is an environment that facilitates positive Energy Balance and weight gain. Our instincts tell us to conserve energy whenever we can.

THE DOWNSIDE OF VALUE-PACKAGING

Successful weight loss depends, in part, on identifying personal, cultural, and societal behaviors that put us at risk of weight gain. Many of these contributing behaviors are integral to our daily activities and we don't normally analyze their impact.

However, let's examine these behaviors. Think about how they play a role in your life. A little conscious effort to review your habits will go a long way toward developing your effective weight management program.

What factors in your environment promote increased energy intake? Make note of them!

There are many possibilities. Is choosing the bigger size—like the seventeen-year-old who chose the super-sized Big Mac meal—something that you sometimes do? Scientific research has shown that the more food you have on your plate, the more calories you are about to ingest.

Look at these statistics[5]:

- Girls and women drank 17 ounces when served a large size soda and only 9 ounces when served a small.
- On average, people ate 165 candy-coated chocolate bits when given a 2-pound bag of the candies and "only" 112 when given a 1-pound bag.
- People eat about 50% more "hedonistic" foods, like candy, chips, crisps, popcorn, when those foods come in bigger packages.
- With other foods (healthier, non-indulgent foods), the increase is usually smaller: only about 25%.

Our culture of consumerism has accepted the increased serving size of many of the most popular foods and beverages. It means we take in more calories. This is true of french fries, soft drinks, and candy bars—in fact, most foods. Some branded food products have more than doubled in portion size over the past few years.

Wherever we go, it seems like the value equation that Americans have adopted is "more for less." How do I get the deal? It doesn't matter whether you're spending three dollars to drive across town to buy candy bars by the box at the warehouse club store, or stopping at the drive-through window of the restaurant. More food for less money is something that seems to be hard-wired into our value circuits, and, for an extra 39 cents, why not?

Junk Food Bargain = Energy Surplus

---

[5] B. Wansink, Director, Food and Brand Research Lab, University of Illinois Champaign-Urbana.

It is not really a surprise that if you are offered more, you will eat more.

In every scientific study measuring consumption, scientists came to the same conclusion—the more people are offered, the more they eat. For instance, Dr. Barbara Rolls and her colleagues found that people ate more macaroni and cheese if there was more on the plate[6].

This is not a conspiracy concocted by the food companies to make you eat more. It is you, the consumer, who demands ever better values. The easiest way for manufacturers to increase perceived value is to increase quantity—and that is exactly what they do.

Increase in Portion Sizes 1955-2001

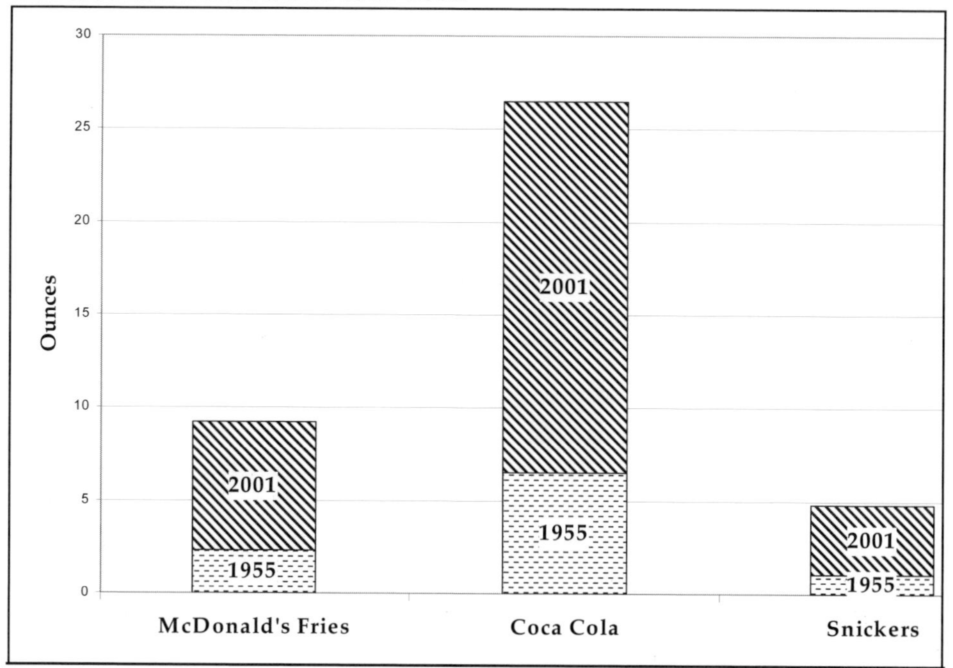

Nutrition Action, 2001.

---

[6] Rolls, B. J., Morris, E. L., & Roe, L. S. Portion size of food affects intake in normal-weight and overweight men and women. *American Journal Clinical Nutrition*, in press.

The graph shows how much package size increased for three popular products between 1955 and 2001. Packages of fries, bottles of soda, and bars of chocolate are now two and three times as big as they used to be. Why?

It is all about getting more product for less money! Creating increased value is part of our culture, part of what our economy strives to accomplish. It is a national habit, but one you must consciously be aware of when it comes to food consumption. More is not necessarily better!

Our food supply provides an abundance of processed foods that are high in fat and high in **energy density**. Energy density is a measure of the number of calories you get in a given weight of food (see Chapter 5 for more on energy density). Pastries, for example, provide a large number of calories in each ounce of food so they have a high energy density. Carrots provide a small number of calories in each ounce and have a low energy density. Because fat has more calories per gram than protein or carbohydrate does, this means that foods high in fat also tend to be high in calorie density.

High Calorie Count/Gram = High Energy Density

A substantial amount of scientific research suggests that, when given high-fat diets, many people take in more total calories than when given low-fat diets. Similarly, when we eat high energy density foods (such as pizza), we eat more than when we eat low energy density foods such as fruits and vegetables. This seems obvious, yet it is important to be very aware of these simple facts.

There are other factors that are important to look at. We also tend to eat a lot of **high glycemic index** foods.

What are high glycemic index foods? These are foods that cause your insulin and blood sugar to increase rapidly after you ingest them. Some experts think these foods will make you hungrier more quickly than low glycemic foods would.

High glycemic foods include white bread, processed breakfast cereals, biscuits or crackers, potatoes, pasta, rice, and tropical fruits such as bananas. Low glycemic foods include whole grain breads, unrefined cereal, legumes, and temperate climate fruits such as apples.

We also get a large number of our calories from beverages such as soft drinks. Many of us may not realize how many calories are in our favorite beverages, or how much we consume. We tend to pay more attention to the calories in food. Extra calories may "sneak in" when we think that all we are doing is quenching our thirst. According to the US Department of Agriculture, the consumption of soft drinks has increased 500% over the past 50 years. USDA estimates for the period between 2000 and 2006 indicate that per capita consumption of soft drinks will be 53 gallons or about 565 cans per year[7].

Let's put this in perspective and see how many calories are sneaking into our diets. A single 12-ounce can of non-diet soda contains 39 grams of sugar—about 10 teaspoons per can. This amount of sugar contains 140 calories and represents 13% of the daily recommended carbohydrate intake based on a diet of 2,000 calories per day.

---

53 gallons of soft drinks each year
= 72.5 pounds of sugar each year

Food and drink is everywhere and it is very inexpensive. It tastes great, comes in endless variety, and is advertised extensively. Is it any wonder that it is hard for many people to push away from the table or say no when there is so much highly palatable food everywhere?

[7] FAS/USDA. *U.S. per capita consumption of specified beverages*. 2001. Available at: www.fas.usda.gov/htp/tropical/2001/06-01/beverage.pdf

OUR CHILDREN ARE AT RISK!

In the United States, at least 11 million children are overweight. This number has tripled over the last 20 years[8]. Many experts say that the likelihood of being overweight correlates to the amount of television your child watches. With the typical child now watching three to five hours of television every day, the risk of childhood obesity is very high.

USDA Food Pyramid

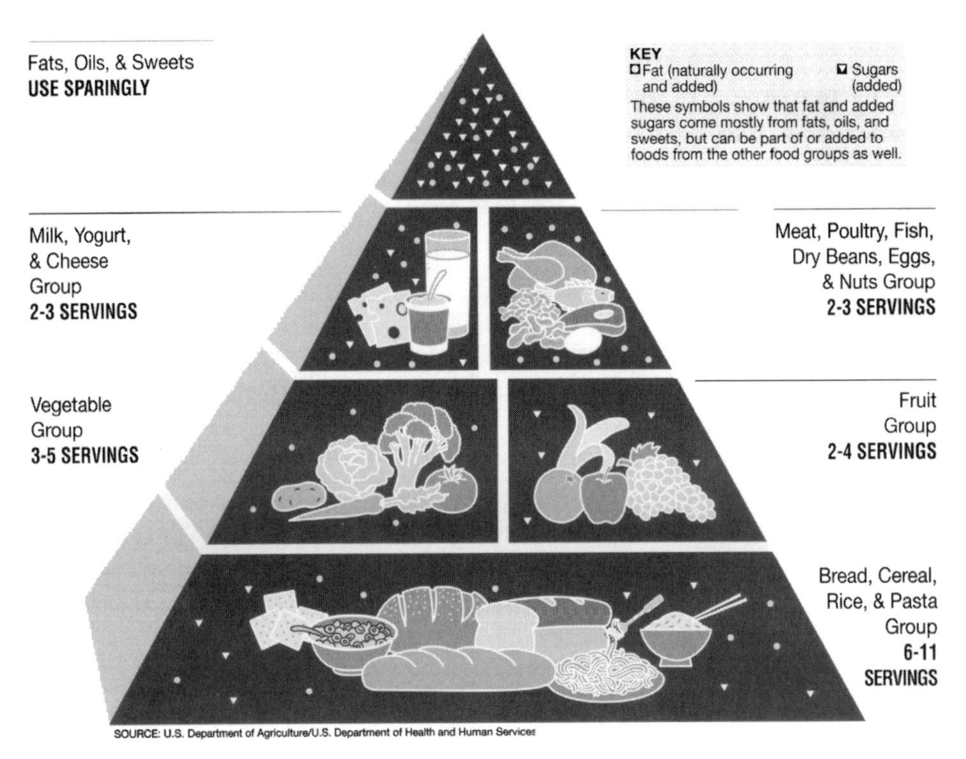

---

[8] U.S. Department of Health and Human Services (2001). *The Surgeon General's call to action to prevent and decrease overweight and obesity*. Rockville, MD: Available from: US GPO, Washington.

Television has a major impact on all of our lifestyles—not just on the lifestyles of our children. Television advertising is incredibly powerful and encourages us to consume too much of foods that should be eaten only sparingly. Children's television programming is loaded with food advertising. Almost all of these advertisements are for products that should be consumed only in limited quantities, such as candy, soft drinks, and potato chips.

These products are included in the top level of the USDA food guide pyramid. Foods at the top of pyramid are those that should be eaten only sparingly.

Sadly, vegetables, fruits, and whole grains—the foods that should constitute the largest part of our diet—don't get much television time. These staples form the foundation of the food pyramid, yet are rarely advertised on television.

Why? It's a simple matter of economics. There's too little profit on commodities like fruits or vegetables to pay for heavy advertising.

WE ARE LESS ACTIVE

If food consumption were the only contributing factor to being overweight, our weight problems would be much easier to solve. Unfortunately, our culture promotes a powerful combination of causal factors. Not only do we eat too much, but both our work and play environments are evolving toward less physical activity.

We have already noted that television influences our activity levels, but it is only one of many behavior shifts over the last few decades that has reduced our individual energy output.

Why walk when you can drive? Why take the stairs when you can take the elevator? Sometimes you can't even find the stairs in a building, let alone use them. Our environment seems to have every labor-saving device imaginable. We even have remote controls built into chairs for operating our television remote controls.

Relatively few jobs these days require high levels of physical activity. Through advances in technology, tasks that used to require physical activity have been automated, and often require no human physical input at all. We have even eliminated physical activity classes in schools. Only one state in the country—Illinois—requires daily physical education. In many schools, recesses have been eliminated or limited to leave more time for academic pursuits.

We have become a nation that relies on labor-saving devices. You might think this would lead to more leisure-time physical activities, but in fact the opposite is true. Remember the previous comments about our ancestors' harsh lifestyles? One of their goals was to reduce physical effort. This has also become a goal of our society—if not consciously, at least as a byproduct of technological advances.

An immense amount of our cultural environment—communications, transportation, drive-through shopping, factory automation—seems to result in reduced physical activity. We buy billions of fast food hamburgers and we pick them up at drive-through windows to avoid climbing out of the car and walking across the parking lot. We make bank deposits, pick up our laundry, fill prescriptions, and get our coffee at drive-through windows.

And what do we do with all the time we saved? We sit.

Most of us don't fill our free time with physical activities. Would you like to sit down at the end of an exhausting day and watch a movie on TV? Or would you rather go for a walk in the park with your family?

If you are part of the minority who choose the walk, then you have already begun establishing behavior patterns that will help you manage your weight. If you typically watch TV, play computer games, and surf the Internet, you are probably being faithful to your genetic heritage and conserving energy.

Relaxing comes naturally to most of us—but it also reduces our physical activity and reinforces our tendency to be overweight.

This may all sound discouraging. But don't give up. Just becoming aware of your activity levels—as you are doing by reading this book— is an important step on your journey to fitness.

Climbing Stairs = Losing 1–2 Pounds

A bit of simple math illustrates the potential impact of small changes in our level of physical activity. Using a television remote control, as compared to walking to the television to change the channel or turn the television on or off, can save enough calories over one year to equal one pound of body fat. Similarly, taking the stairs at work instead of the elevator could result in the loss of one or two pounds of body fat each year.

Think of all of the labor-saving devices you use each day! Each one reduces your energy expenditure—and that saved energy is eligible to be stored as fat.

The table below shows the calories that were saved (that is, not used up) by reliance on technology. You see that, every year, these labor-saving devices may be making you five pounds heavier.

Energy savings per year from common labor-saving devices

Calorie saving activity	Calories saved Female	Calories saved Male
Reliance on elevators and escalators	4,160	4,485
Reliance on TV remote control	728	819
Decrease in work-related activity	13,000	13,000
Totals about 50 calories per day	Total saved: 17,888 calories	Total saved: 18,304 calories
Equivalent to about 5.1 pounds of body mass		

Authors

We should also look at how the structure of our communities affects our levels of physical activity. In traditionally-structured communities, things were laid out in a grid pattern. You could walk to school, you could walk to the ball field, and you could walk to the store. It was all very safe and easy to get to.

Today, however, few suburban neighborhoods are laid out using the traditional grid design. Instead, we have urban sprawl and cul-de-sacs. You may not feel you can safely walk so you drive. If you want to go to the store, you may have to cross multiple lanes of major traffic and there may not be crosswalks or sidewalks.

Our changing communities have altered the fabric of our lives in ways that were not intended. We did not set out to design communities that would encourage us to gain weight, but that is exactly what has happened.

SUCCESS HAS ITS DRAWBACKS

In a way we could say that obesity is the result of societal success. Most of us grew up in an environment where it was a desirable thing to have abundant, accessible, affordable food.

Think about your parents, your grandparents, your great-grandparents. Abundance was something they dreamed of, because they had to work so hard to get what they had. Likewise, everyone wanted to reduce the amount of physical labor required for daily living. People had to work eight, ten, or even sixteen hours a day just to make ends meet.

Long work hours also put a premium on free time—time that was used to recuperate from the demands of physical labor. Everybody wanted a little free time. They wanted to be able to choose how they spent that free time, and they wanted to live the good life.

That goal was passed from generation to generation. We want a better life for our kids, just like our parents wanted a better life for us.

There are socio-cultural drivers at work here. How do you achieve the good life?

The American Dream + Survival Instincts = Weight Gain

The good life results from increased productivity. That's what makes society go around. The desire for increased productivity leads to a greater demand for technology. To increase productivity, we often dedicate more time to work. Our commitments seem to multiply.

Less free time increases our demand for convenience. Many of you do not even have time to go home and cook dinner after work. So fast food restaurants have sprung up to meet your need for quick, convenient food.

Convenience, comfort food, less physical activity, and more sedentary leisure-time activities are all consequences of seeking the good life. Weight gain and obesity are unintended consequences of our having succeeded at everything we set out to accomplish as a society a hundred years ago.

We now live in an environment that exerts constant pressure on us to eat more and be less physically active. Given this situation, it is surprising that everyone is not obese.

This is not to say the situation is hopeless. The fact that not everyone is obese suggests that some people **can** manage their weight despite the environment.

It may be useful to consider how these people maintain their healthy weight in an environment that promotes weight gain.

WHAT TO DO

To succeed with weight management, look first at your own environment—at your household, your family, your friends, and your workplace. Then go beyond your immediate environment to your neighborhood and community to identify the factors that are affecting your behavior.

When you take control of those factors, you begin to take control of your life and your weight. The solution is not to transport ourselves back to the nineteenth century, when the harsh environment ensured that obesity was not a general problem. We are never going to reject technology and all the good things that have come from it. Rather, we need to recognize the behaviors that translate to a positive Energy Balance and learn how to manage them.

What tactics and tips are needed to do that? **The most important element is something called cognitive management of our behavior.** That means you have to think about your lifestyle. In our current environment, we can no longer rely on our instincts to tell us when to eat and when to rest.

To appropriately modify these instinctive behaviors takes substantial effort. You have to counter your natural tendency to eat lots and move little. You need to think about the sum of your activities around the clock, not just during the thirty minutes you spend practicing a formal exercise program. What modest changes could you make in your daily regime that would increase your level of physical activity?

It really is about mind over matter. This is where the tools provided in this book can help us manage the complexity of our behaviors and enable us to successfully manage our weight even though we are in an environment that discourages healthful habits.

Think about your lifestyle!

CHAPTER 3. WHY WORRY ABOUT WEIGHT GAIN?

Holly Wyatt, M.D.

In previous chapters, we noted that 65% of Americans are overweight and that the average American gains one to three pounds every year. This means that less than 35% of us weigh the amount that is right for our bodies. And this number is likely getting smaller each year.

In this chapter, we will consider the consequences of this "fattening of America." What are our options? One is just to accept the fact that we are getting bigger and change our environment to accommodate larger people. The second—far wiser—option is to recognize that the unrelenting weight gain that has engulfed our society is a potentially severe problem and that it is associated with substantial health risks.

The next step is to do something about it.

OVERWEIGHT OR OBESE?

Let's begin by defining overweight and obesity.

Some of you know you are overweight. The clothes you buy keep getting bigger and you know you need to lose a few pounds. Others among you may be thinking that you have gained a few pounds, but you're not sure if you fall into that 65% of Americans who are overweight.

The widely accepted way to evaluate overweight is by using body mass index or BMI. The medical community no longer uses the term "ideal body weight" in discussing weight problems. The preferred term now is "healthy weight." Healthy weight is defined as BMI between 18.5 and 25 kg/m^2.

18.5–25 BMI = Healthy Weight

BMI is a calculation based on body weight and height. The chart on page 57 allows you to calculate your BMI from pounds and inches (see Components of Weight Loss in Appendix A for additional information).

While BMI does not directly measure body fat, and it is really excessive body fat rather than excess of other body tissue that creates a health risk, BMI is highly correlated with body fat. Accurate techniques to measure body fat directly are not easily available or affordable for widespread use. BMI is easy to calculate and serves as a standard measure and a strong indicator of body fat. The *SANY* website contains an easy to use BMI calculator that can be found at; http://www.shapinganewyou.com.

BMI, while not perfect as a measurement of fat, is a valuable tool for identifying individuals whose weight has increased to the point where it can negatively impact their health. One advantage of BMI is that it allows people of various heights and weights to be compared with each other. If you are tall, the weight at which you are considered overweight is higher than it would be for someone shorter.

You can think of your BMI as one of those important numbers that tells you about your health. It is as valuable as your cholesterol level or your blood pressure. More and more physicians are treating BMI as a vital sign.

BMI in pounds and inches

BMI	19	20	21	22	23	24	25	26	27	28	29	30	35	40
Height (in.)	Weight (lb.)													
58	91	96	100	105	110	115	119	124	129	134	138	143	167	191
59	94	99	104	109	114	119	124	128	133	138	143	148	173	198
60	97	102	107	112	118	123	128	133	138	143	148	153	179	204
61	100	106	111	116	122	127	132	137	143	148	153	158	185	211
62	104	109	115	120	126	131	136	142	147	153	158	164	191	218
63	107	113	118	124	130	135	141	146	152	158	163	169	197	225
64	110	116	122	128	134	140	145	151	157	163	169	174	204	232
65	114	120	126	132	138	144	150	156	162	168	174	180	210	240
66	118	124	130	136	142	148	155	161	167	173	179	186	216	247
67	121	127	134	140	146	153	159	166	172	178	185	191	223	255
68	125	131	138	144	151	158	164	171	177	184	190	197	230	262
69	128	135	142	149	155	162	169	176	182	189	196	203	236	270
70	132	139	146	153	160	167	174	181	188	195	202	207	243	278
71	136	143	150	157	165	172	179	186	193	200	208	215	250	286
72	140	147	154	162	169	177	184	191	199	206	213	221	258	294
73	144	151	159	166	174	182	189	197	204	212	219	227	265	302
74	148	155	163	171	179	186	194	202	210	218	225	233	272	311
75	152	160	168	176	184	192	200	208	216	224	232	240	279	319
76	156	164	172	180	189	197	205	213	221	230	238	246	287	328

Source: www.shapinganewyou.com

Now is a good time for you to determine your BMI. You need to know your height without shoes and your weight in light clothing.

Calculate your BMI from the BMI chart. Find your height in inches along the left side and then move to the right to find the weight in pounds that is closest to the amount you weigh. Your BMI is above your weight, in the top row.

Record your BMI in a journal or a weight management record such as the one found on page 165. Keeping a journal is very important. We will say more about journals later.

Once you have your BMI, how do you tell what it means?

The chart above will show you whether you are at a healthy BMI, or whether you are overweight or obese. If your BMI is at least 18.5 and less than 25, you are in the healthy weight range. If your BMI is less than 18.5, it does not necessarily mean your weight is too low, but some people below 18.5 can improve their health by gaining weight.

BMI> 25 can mean high health risk.

If your BMI is 25 or greater but less than 30, you are in the overweight category. Being overweight does increase disease risk but not as much as if your BMI is over 30.

We classify those with BMI 30 or above as obese and it is this category where weight is most likely to negatively affect health.

THE IMPORTANCE OF YOUR WAIST MEASUREMENT

Your waist circumference is another important indicator of health risk[9]. The bigger your waist, the more fat you have stored around your middle. It is this extra fat around the middle that seems to most frequently lead to more serious conditions such as diabetes and heart disease.

Waist circumference can be used in combination with BMI to help determine the extent to which your weight negatively contributes to your health. The medical community used to evaluate the waist-to-hip-ratio (WHR), that is, the circumference of the waist divided by the circumference of the hip. But we now know that a simple waist measurement provides more information about fat around the middle than the WHR.

---

[9] National Institutes of Health: Clinical guidelines on the identification, evaluation, and treatment of overweight and obesity in adults – the Evidence Report. *Obesity Research*. 1998;6: (Supplement 2) 51S-210S.

Levels of risk

Risk of associated disease based on BMI and waist size			
BMI	Designation	Waist less than or equal to 40 in. (men) or 35 in. (women)	Waist greater than 40 in. (men) or 35 in. (women)
18.5 or less	Below normal	—	N/A
18.5–24.9	Normal	—	N/A
25.0–29.9	Overweight	Increased	High
30.0–34.9	Obese, Class 1	High	Very high
35.0–39.9	Obese, Class 2	Very high	Very high
40+	Obese, Class 3	Extremely high	Extremely high

NIH

Here's why the waist measurement is important. If a person has a lot of muscle, the extra muscle may be reflected in a high BMI. In other words, many professional athletes have a high BMI even though their body fat content is low. These athletes' low waist circumference is the signal that tells us that body fat content is low despite the high BMI.

Most overweight or obese people will have both a high BMI and a high waist circumference. A low waist circumference with a BMI in the overweight category might suggest greater muscle mass rather than greater fat content. Experts consider a waist circumference greater than 40 inches in men and 35 inches in women to be high.

BMI > 30 + Large Waist = Extreme Health Risk

It is important to measure your waist in the right place. Your waist circumference is not necessarily your pant size. Many people wear their pants below their waist. Your waist size is the circumference measured where your waist is largest.

One way to find this place is to find the top of your hip bone on the right and on the left. Put the tape measure around your waist at this level to find your waist circumference. Have a look at the diagram below, and take your measurement now. Record this along with your BMI in the weight management log on page 165.

Measuring your waist

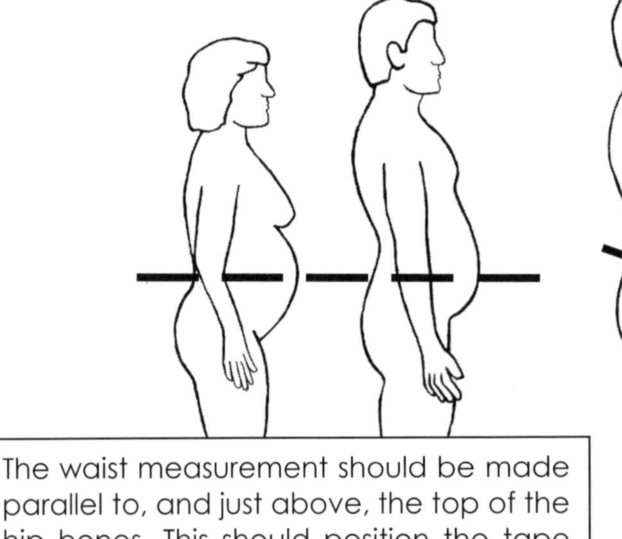

The waist measurement should be made parallel to, and just above, the top of the hip bones. This should position the tape at, or just below, your belly button. Take the measurement after you exhale and just before you begin another breath.

WHAT ARE THE CONSEQUENCES?

Let's get back to whether or not we should worry about the increasing number of overweight people in the United States. Since 65% of Americans are overweight, perhaps we should simply modify our environment to better accommodate larger people. Let's just get bigger clothes, bigger chairs, bigger cars, and so on, and go about our business.

Before we decide to do this, it is important to understand some of the consequences of excess body fat. It has been proven that carrying extra weight and extra body fat has many serious health-related consequences. In fact, overweight and obesity are associated with increased risk for most of the chronic and potentially deadly diseases in the United States.

The list of diseases excess body weight makes you vulnerable to is frightening. Have a look…

- Being overweight or obese dramatically increases the probability that you will get type 2 or adult-onset diabetes. This is a very serious disease that can lead to kidney disease, amputation, and heart disease.
- Being overweight or obese directly increases the chances that you will have heart-related problems such as high blood pressure, high cholesterol, heart attacks, or stroke. As you may know, heart disease is the number one killer of both men and women in the United States.
- Excess body weight can lead to other hormonal problems, such as infertility and irregular periods in women.
- Excess body weight can cause an excess of uric acid, leading to joint pain or gout.
- Excess body weight can lead to other musculoskeletal problems that cause joint pain or even arthritis.
- Some cancers, such as breast, colon, prostate, and uterine cancer have been related to carrying extra weight.
- Being overweight or obese is associated with gall bladder disease, liver inflammation, and stomach problems.
- Excess body weight is very closely associated with sleep apnea. Because people with sleep apnea stop breathing for periods of time at night, they don't ever get a really good night's sleep. This can place immense stress on the heart.
- Being overweight or obese is associated with increased risk of psychological problems, such as depression or poor self-image.

The list seems nearly endless. Clearly, being overweight or obese puts you in danger of developing a very serious medical problem.

You may be overweight or obese now and may not be experiencing any of these health problems. Does that mean you don't have to worry?

Well, quite frankly, the odds are lousy. The higher your BMI, the greater your risks of contracting one of the diseases mentioned above.

If your BMI is 35...
You are 40 times more likely to develop diabetes than if your BMI were 22.

If your BMI is 40...
You are 100 times more likely to develop diabetes than if your BMI were 22.

Do you remember what your BMI is? If you've forgotten or you haven't yet calculated it, turn back to page 57 now and use the chart to look up your BMI. Then note your position on the risk chart on page 59.

Where do you fit on the health risk continuum?

DISPROPORTIONATE HEALTH RISKS

We have seen that having extra body weight makes it more likely that you will develop a serious medical condition, such as diabetes, in the future. This is true even if you are perfectly healthy now except for having extra weight.

It's all about probability. Not every overweight or obese person will develop a serious medical problem—but extra body fat increases the probability enormously. The more overweight you are the more likely you are to have these health problems.

The next table shows what proportion of the incidence of some of the diseases we've talked about is due to obesity. In two diseases—sleep apnea and type 2 diabetes—the majority of cases are due to obesity.

Obesity is a factor

Proportion of disease due to obesity	
Sleep apnea[5]	95%
Type 2 diabetes[1]	61%
Gallbladder disease[1]	30%
Asthma[4]	25%
Hypertension[2]	25%
Coronary heart disease[1]	17%
Osteoarthritis[3]	14%
Breast cancer[1]	11%
Uterine cancer[1]	11%
Colon cancer[1]	11%

1. Wolf, A. M., & Colditz, G. A. *Obesity Research.* 1998;6:97.
2. American Health Foundation Roundtable on Healthy Weight. *American Journal of Clinical Nutrition.* 1996;63(supplement):4095.
3. Camargo, C. A., Jr., et al. *Archives of Internal Medicine.* 1999;159:2582.
4. Gelber A. C., et al. *American Journal of Medicine [JAMA}.* 1999;107:542.
5. Suratt, P. M., & Findley, L. N. *New England Journal of Medicine.* 1999;340:881.

IMPROVE YOUR ODDS

What if you are one of the unlucky people who already have type 2 diabetes or another condition associated with being overweight? Should you do something about your weight?

The answer to this question is an unequivocal **yes**. If, for example, you have type 2 diabetes, your weight helps determine the effectiveness of your treatment. Losing weight can mean that you will require less medication or that the medication you take will be more effective in controlling your disease. Even a small weight loss will make it much easier to control your disease.

Even a small weight loss makes a big difference!

Additionally, if you already have one weight-related condition, you are at risk for developing another such condition. Controlling your weight can help prevent this from happening.

Even if you do not develop any of the diseases that are weight-related, having a BMI over 30 is associated with a greatly increased risk of death. Life insurance companies have known this for a long time—they often charge overweight people higher premiums for life insurance.

BMI > 30 = Danger!

Why? Because when you are overweight, you have a 50% to 100% greater risk of dying—from any cause—than someone with a BMI in the healthy range. That's right—your risk of dying might now be double what it will be when you reach your healthy weight!

Let's consider the most frequent causes of preventable death in the United States. Smoking cigarettes is the number one cause, but obesity and physical inactivity come in second. It is estimated that over 300,000 people die each year in the United States from causes related to obesity and physical inactivity.

Too much weight + too little physical activity = 300,000 preventable deaths every year!

Remember, we are talking about preventable death. Everyone can greatly reduce their chances of premature death by keeping their weight under control and getting regular physical activity.

The graph below vividly illustrates the danger of being overweight or obese. You can see that carrying extra body fat has caused twice as many preventable deaths as alcohol, six times as many as guns, and about fourteen times as many as motor vehicles.

These statistics are not pleasant, but they give us persuasive reasons to manage our weight problems as quickly and effectively as possible.

Causes of Preventable Death in the United States

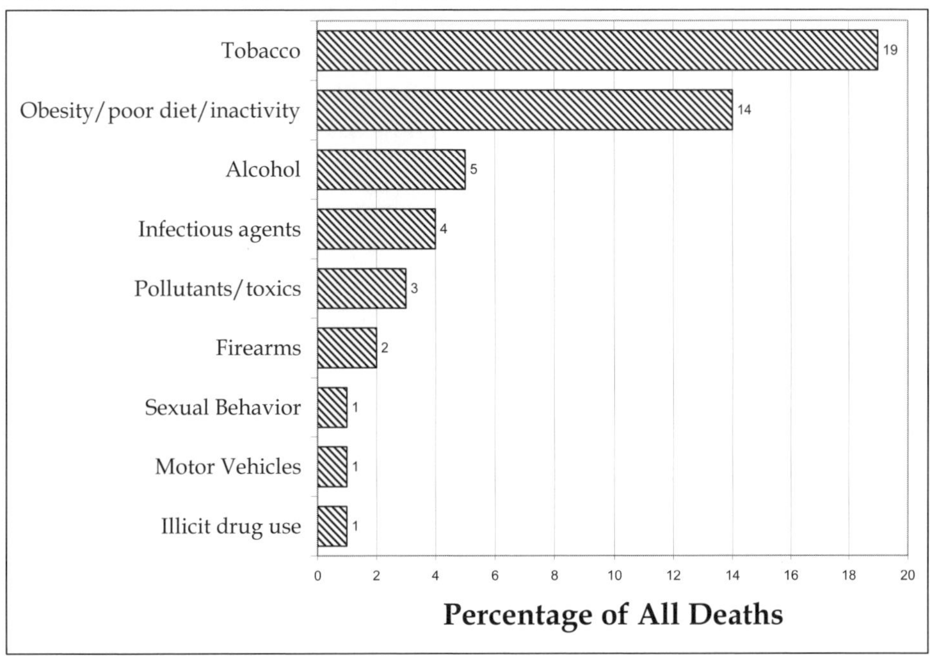

Source: McGinnis, J.M., & Foege, W. H. Actual causes of death in the United States. JAMA. 1993; 270:2207-12 (1990 data).

Let's face it—if you are obese, your life is in danger. To be healthy, you need to pay attention to your body weight. It is true that being overweight does not guarantee that a person will fall victim to a weight-related disease—but it certainly increases the odds!

Even if you beat the odds and stay physically healthy, you are still at risk. Being overweight is associated with a high risk of developing psychological problems. This may be related to the fact that overweight people are stigmatized and discriminated against in our society. This is something overweight people have to face every day—extra weight is visible and everyone can see you have that health problem. People may not know that you have diabetes or heart disease, but it is impossible to hide too much weight.

People who have a weight problem are exposed to public disapproval. In a perfect society this would not be the case, but the reality is that being overweight is one of the least socially accepted forms of visible difference. The effects of this can be seen in education, employment and health care.

For example, statistically speaking, young women who are obese do not reach as high a level of education as those of normal weight. Overweight or obese women are less likely to get married and are more likely to have such low household income that they fall into the poverty category[10].

Not all overweight or obese persons have psychological problems, but many do. The psychological difficulties vary from individual to individual. Some experience intense suffering; these people are deeply unhappy about being overweight. They feel shame and guilt and have a poor self-image. Their daily life is strongly affected and their overall quality of life suffers as a result.

You do have control!

On the other hand, individuals who successfully lose weight and keep it off consistently report that their quality of life is better after weight loss than before[11].

Where does this leave us? The bad news is that being overweight can negatively affect your health and the quality of your life. The good news is that you can markedly decrease your health risk and greatly improve the quality of your life.

How? Lose weight and keep it off.

---

[10] Gortmaker, A., Must, A., Perrin, J. M., Sobol, A. M., and Dietz, W. H. Social and economic consequences of overweight in adolescents and young adulthood. *New England Journal of Medicine*. 329:1008-1012, 1993.

[11] Wing, R. R., & Hill, J. O. Successful Weight Loss Maintenance. *Annual Review of Nutrition 2001*; 21:323-341.

If you are overweight, there **is** something you can do about it. You are not doomed to develop a serious medical condition. You are not doomed to suffer endlessly from unfair discrimination.

You can actually make a change. You can do something about it. You do have control.

How much weight loss is required to see improvement in your health and quality of life?

This is a very important question and there has been a great deal of scientific research on this point. Research consistently shows that very large health benefits can be achieved through relatively modest weight loss.

This is good news! You do not have to get back to the healthy BMI range to significantly improve your health and quality of life. This is particularly important for individuals who already have high BMIs.

The summary below lists some of the very positive health effects of even a small weight loss. Wouldn't you agree these health benefits are worth having?

Health benefits of modest weight loss:

- Decreased blood pressure.
- Decreased risk of heart attack or stroke.
- Decreased blood glucose and insulin levels.
- Decreased LDL ("bad" cholesterol) and triglycerides.
- Increased HDL ("good" cholesterol).
- Reduced symptoms of degenerative joint disease.
- Improvement in gynecologic conditions.

For example, if your BMI is 40, you would have to lose a lot of pounds to get to a BMI of 25. If you were 5′8″ tall and weighed 262 pounds, you would need to lose 98 pounds to go from a BMI of 40 to a BMI of 25.

It takes time and dedication to lose that amount of weight. It could take years. But fortunately, even losing as little as 5% to 10% of your weight can lead to significant improvement in your risk of disease and your quality of life. For example, if you weigh 200 pounds now, you can get important positive results by losing between 10 and 20 pounds. Most people are able to lose this amount of weight and keep it off.

We do not mean to say that you have to stop when you have lost 5% to 10% of your weight—many people lose more than that. The more you lose, the more you improve your health and quality of life. Just remember—even a little bit of success in weight loss can lead to big improvements for you.

Small Weight Loss = Huge Health Gain

A 5% to 10% loss of body weight for someone who is overweight or obese decreases their risk of heart attack or stroke. It can decrease your blood glucose levels, your blood pressure, and your cholesterol levels. This small weight loss can dramatically improve sleep apnea, making it disappear or at least making it easier to treat. Modest weight loss can substantially reduce joint pain and arthritis. In other words, you get a large benefit from a small weight loss.

Why does modest weight loss produce such a big improvement in your health? It seems to be because, when you start losing weight, much of the first weight you lose is fat and much of the fat comes from around your middle.

Remember, this is the fat that scientists think causes most of the metabolic problems associated with being overweight or obese. Even a small weight loss, in the range of 5% to 10%, will significantly reduce the amount of fat you're carrying in your waist area.

The National Institutes of Health recently completed a very large study demonstrating the power of small weight loss to improve health.

This study, called the Diabetes Prevention Program (DPP)[12] studied 4,000 subjects who were overweight. The subjects were considered to be at high risk of developing type 2 diabetes because their bodies had begun showing resistance to insulin. These people did not yet have diabetes, but it was likely that many of them would develop the disease in the future.

The intent of the study was to see if modest weight loss could reduce the risk of diabetes in this highly susceptible group of people. The 4,000 subjects were randomly assigned to groups and tracked for three to six years.

One of the treatment groups was a lifestyle change group where the aims were to reduce body weight by 7% and increase physical activity by 150 minutes per week. In fact, these subjects did lose 7% of their weight but they gained 2% back. In other words, they maintained a 5% weight loss. They lost weight by eating a low-fat diet, watching total calories and increasing physical activity.

A second group was given a drug (Metformin) and a third group did not change what they were already doing. A large number of the people who did not make changes developed type 2 diabetes over the next three to six years. The group given Metformin had about 33% less diabetes, but the greatest benefit was in the lifestyle group. The lifestyle group reduced the risk of getting diabetes by almost 60%.

This shows that, with modest weight loss and modest increase in physical activity, the risk of developing diabetes was cut by more than half. This is a powerful result.

The graph below illustrates the difference between the two groups. Clearly, lifestyle change was far more effective than pharmaceutical treatment at reducing the risk of diabetes.

Let's look at what we've discovered so far.

Just accepting that we are all getting fatter is not a good solution to the problem. The close association of overweight and obesity with serious

---

[12] National Institute of Diabetes and Digestive and Kidney Diseases (NIDDK) (2001). *Diet and Exercise Dramatically Delay Type 2 Diabetes: Diabetes Medication Metformin Also Effective.* Bethesda, MD.

diseases such as diabetes and heart disease suggests that this strategy would eventually overwhelm our health care system.

Even if you are not concerned about the health risks of excess weight, you have a greater chance of suffering from discrimination, psychological problems, and decreased quality of life if you are obese. Just accepting obesity is not a good decision.

Lifestyle changes work better to reduce risk

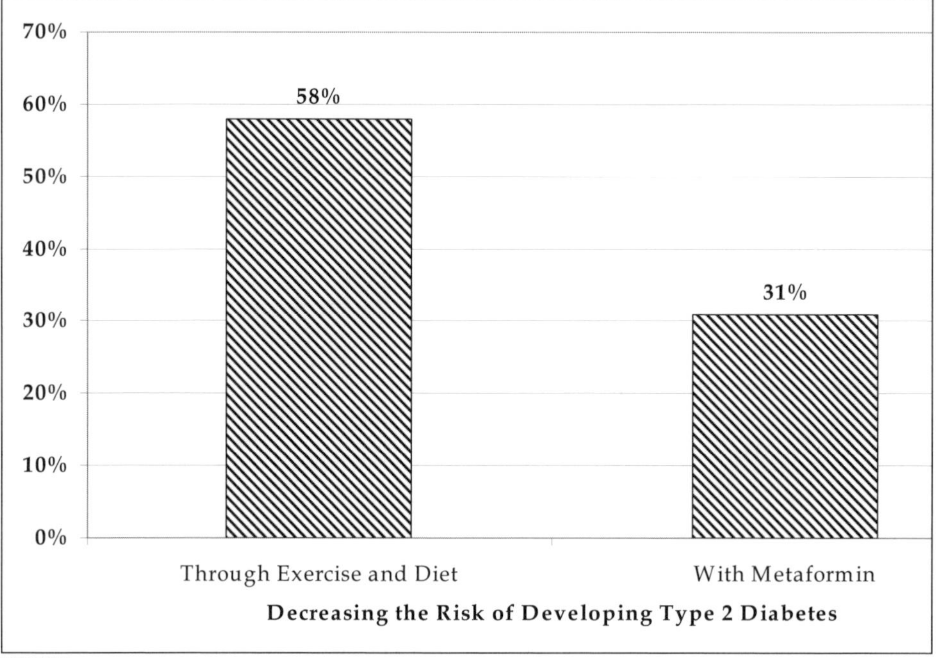

NIDDK

While the effects of obesity are serious, the good news is that you can do something about your excess body weight. Simply losing modest amounts of body weight can have powerful effects on your health and on your quality of life. Almost everyone can make enough lifestyle change to achieve modest weight loss. The tools in this book and the *SANY* software will help you change your lifestyle and produce weight loss that can improve your health and your entire outlook on life.

CHAPTER 4. THE BASICS OF ENERGY BALANCE

Holly Wyatt, M.D.

In this chapter we will go into detail about Energy Balance. You have now heard this term several times. We want to make sure you understand the basics of Energy Balance, because only then will you be able to achieve your weight management goals.

Changing your body weight is about changing your Energy Balance—there is simply no other way. We want to make Energy Balance real for you and show you how to use it to set and achieve your goals.

Your body weight is not going to magically disappear. The only way to lose a pound of body weight is to accumulate an energy deficit of about 3,500 calories. Similarly, the only way to keep from regaining the weight you lose is to ensure that your energy intake and energy expenditure are equal. When this is the case, it is impossible for you to gain weight.

Let's relate Energy Balance to a concept you may be familiar with—balancing your checkbook. The skills needed to succeed in balancing your checkbook are similar to the skills needed to manage your weight.

If you keep track of your bank deposits and the interest you earn and also keep track of the checks you write, you know your checkbook will balance exactly—at any time. Money does not magically appear in, or disappear from, your account. There are no surprises.

Similarly, if you know exactly how many calories you take in and exactly how many calories your body burns, you know your body weight exactly—at any time. There will be no surprises when you step on the scale.

Many people have been brain-washed by fad diets and other sources of "expert" opinion which tell them to forget about counting calories.

We disagree. Understanding what goes into your body is very important and well worth the effort it takes to keep track. You would not expect to maintain fiscal health without keeping track of your expenses, cash flow, and assets. Is your health any less important than your bank account?

You balance your checkbook in dollars—dollars you deposit, dollars you spend, and dollars you have in your account.

Your energy "checkbook" is the same way. You balance your energy in calories—calories you eat, calories you burn, and calories you have in your weight account (stored in your body).

Think of calories as dollars. The principles for balancing your checking account and balancing your energy "account" are very similar. The goals are different—you want to gain money and lose weight—but the accounting processes are the same.

To review, Energy Balance is the concept of balancing the energy (calories) in the food you eat with the energy (calories) your body burns. If energy intake exceeds energy expenditure, you are in positive Energy Balance (more coming in than going out) and you will gain weight. You will gain approximately 1 pound for each 3,500 calories of positive Energy Balance.

Conversely, if your energy expenditure exceeds your energy intake, you are in negative Energy Balance (more going out than coming in). You will lose approximately 1 pound for every 3,500 calories of negative Energy Balance.

Finally, if your energy intake is the same as your energy expenditure, you are in zero Energy Balance. You cannot (no matter what!) gain or lose weight if you are in zero Energy Balance.

Your strategy is straightforward. You want to create negative Energy Balance until you have lost the desired amount of body weight. After that you want to create zero Energy Balance and maintain it for the rest of your life.

It sounds very simple put this way, but this really is your goal. We simply have to deal with the details of how to achieve your goal. Let's take a closer look at how your body handles energy.

ENERGY INTAKE

You consume food, and food has energy. Food energy is measured in calories (often written as kCal) and food quantity is measured in grams. Fat is the most calorically dense food, providing 9 calories for each gram you consume. Protein and carbohydrate each provide 4 calories per gram consumed. Weight for weight, fat provides more than twice the calories that proteins or carbohydrates provide.

Because most of us in the United States are less familiar with the use of grams as a weight measurement than we are with ounces, it might be difficult to associate these numbers with a meaningful value.

So what if a gram of fat has 9 calories? Nine isn't such a big number.

See if you feel differently after reading the conversion values. A king-size order of Burger King fries weighs between 6 and 7 ounces, or 190 grams. It contains 600 calories, of which 270 come from fat. Those 9 calories per gram add up in a hurry.

Fat...

Is the densest food, with 9 calories per gram[13] or 254 calories per ounce.

Alcohol...

Has 7 calories per gram, or 198 calories per ounce.

Proteins and carbohydrates...

Have 4 calories per gram, or 113 calories per ounce.

---

[13] An ounce is equivalent to 28.3 grams.

If you had very accurate information about the weight and composition of the food you eat, calculating your energy intake would be easy. You'd be able to determine precisely the number of calories you were taking in. The values highlighted in the box above will give you an approximate guide. As you see, fat has the most calories of the four groups while proteins and carbohydrates have the least.

Since most people do not weigh all of the food they eat, and since many foods are mixed in the sense that they contain a mixture of fat, carbohydrate and protein, it is difficult to determine the precise number of calories you take in. However, it is possible to estimate caloric intake.

The Shaping a New You (*SANY*) software contains over 16,000 listings of caloric values and can calculate your intake very precisely. There are a number of excellent books available that also provide caloric values for thousands of foods and food combinations.

If you are not using the *SANY* software, you will find the log form provided in Part II useful for keeping track of your food intake. This will give you an idea of the number and type of calories you now consume and will help you determine how much you need to change your food intake to achieve your weight loss goal.

ENERGY EXPENDITURE

Your body uses calories for fuel to meet its energy needs. Just as your car needs gasoline to function, your body needs calories.

Your body needs energy for many purposes—to maintain your body temperature, to keep your heart pumping and your lungs functioning, to continually manufacture and dispose of hormones.

Because your body is constantly functioning, it is always burning calories—even while you are sitting, lying down, or even sleeping. This slow rate of energy expenditure is called your **resting metabolic rate** or RMR. When you perform work (physical activity), the rate at which your body uses calories increases.

This is just like what happens in your car's engine. When you want the car to go faster, you apply pressure to the gas pedal and your car uses more fuel as it increases its speed.

RMR = Energy Out At Rest

Like your car, your body uses more energy when you increase your physical activity. The total number of calories your body burns is the sum of your RMR plus the energy expended in physical activity. The more intense your activity per unit of time, the greater the number of calories used as fuel.

RESTING METABOLIC RATE (RMR)

Your RMR is determined primarily by the size of your body. Larger bodies have higher RMRs, just as larger cars usually have a higher rate of fuel consumption. The easiest way to find your RMR is to visit the Shaping a New You website (http://www.shapinganewyou.com) and go to the calculator page in the Smart Move section. You could also manually calculate your RMR, but the calculation can be cumbersome and much easier to acquire off the Internet—or you can have it truly personalized by using the *SANY* software.

So long as your body weight remains constant, your RMR stays the same. When you gain weight, your RMR goes up and when you lose weight, your RMR goes down.

Many people notice that they lose weight much faster at the beginning of a diet than later on. This is because as you lose weight, your RMR goes down, reducing your negative Energy Balance.

In this sense, your body works against you in losing weight. But since you now know your RMR will go down as you lose weight, you will be able to compensate for this when you develop your weight loss plan.

Weight Loss + Zero Increase in Exercise = Lower RMR

Another factor that contributes to your energy output is something known as the **Thermic Effect of Food**, or TEF. TEF refers to the energy your body uses to digest and process your food. It is equivalent to about 8% to 10% of the energy you take in. For example, if you eat 2,000 calories a day, you burn about 200 calories in TEF. But if you reduce your energy intake to 1,500 calories, your TEF will go down to only about 150 calories per day.

ENERGY EXPENDED IN PHYSICAL ACTIVITY

Your physical activity level is a major determinant of how much energy you expend. The more physically active you are, the more energy your body burns.

The table below shows the energy consumption for one hour of different types of physical activities. It will show you how much you can increase your total energy expenditure by doing different types and amounts of physical activity. This table uses energy expenditure rates that include a factor for RMR based on the averages for various weight levels. Part II explains the processes of calculating energy expenditure in more detail. Care must be taken when using various compilations for energy consumption to know if the RMR is included in the value so that it is not double counted.

Traditional estimates of energy expended in physical activity

Exercise	Weight levels				
(Values are for 1 hour of exercise)	**125**	**150**	**175**	**200**	**250**
Aerobics, general	340	408	477	545	681
Aerobics, low impact	284	340	397	454	567
Backpacking, general	397	477	556	635	794
Basketball, non-game, general	340	408	477	545	681
Basketball, shooting baskets	256	307	358	409	512
Bicycling, <10mph, leisure	227	272	318	363	454
Bicycling, 10–11.9mph, light effort	340	408	477	545	681
Bicycling, 14–15.9mph, vigorous effort	567	681	794	908	1135
Bicycling, stationary, general	284	340	397	454	567

(Values are for 1 hour of exercise)	125	150	175	200	250
Billiards	142	171	199	228	285
Bowling	170	204	238	272	340
Calisthenics (pushups, sit-ups), vigorous	454	545	635	726	908
Calisthenics, home, light/moderate effort	256	307	358	409	512
Child care: sitting/kneeling-dressing	170	204	238	272	340
Cleaning, house, general	199	239	279	318	398
Cooking or food preparation	142	171	199	228	285
Dancing, general	256	307	358	409	512
Gardening, general	284	340	397	454	567
Golf, general	227	272	318	363	454
Health club exercise, general	313	375	438	500	625
Hiking, cross country	340	408	477	545	681
Jogging, general	397	477	556	635	794
Mowing lawn, general	313	375	438	500	625
Pushing or pulling stroller with child	142	171	199	228	285
Raking lawn	227	272	318	363	454
Rope jumping, moderate, general	567	681	794	908	1135
Rowing, stationary, light effort	539	647	755	863	1079
Running, general	454	545	635	726	908
Shoveling snow, by hand	340	408	477	545	681
Soccer, casual, general	397	477	556	635	794
Softball or baseball, fast or slow pitch	284	340	397	454	567
Stair-treadmill ergometer, general	340	408	477	545	681
Stretching, hatha yoga	227	272	318	363	454
Sweeping garage, sidewalk	227	272	318	363	454
Swimming laps, freestyle, moderate effort	454	545	635	726	908
Swimming, leisurely, general	340	408	477	545	681
Tennis, general	397	477	556	635	794
Volleyball, noncompetitive, team (6–9)	170	204	238	272	340
Walk/run-playing with child(ren), moderate	227	272	318	363	454
Walking, 3 mph, mod. pace, walking dog	199	239	279	318	398
Walking, 4 mph, very brisk pace	227	272	318	363	454
Water aerobics, water calisthenics	227	272	318	363	454
Weight lifting, light or moderate effort	170	204	238	272	340

Find the activities you enjoy most and see how much energy they expend. Is there another activity you also enjoy that consumes more energy—and thus increases your energy output?

INDIVIDUAL DIFFERENCES IN ENERGY EXPENDITURE

Even after we account for size, not everyone's body burns calories at the same rate. Let's go back to the car analogy. Different car engines have different fuel efficiencies. Some burn fuel more quickly than others. How quickly they burn the fuel depends, in part, on the quality of fuel and on whether the engine is well-tuned.

The same is true for bodies. Some bodies burn calories at a higher rate than others do, even if they are exactly the same size. That's why we can only estimate your RMR based on your size.

We can't tell you exactly what your RMR is unless we take advantage of technology, as the *SANY* software allows us to do. Your RMR might be higher or lower than the estimate based on your body weight, simply because your body burns calories at a higher or lower rate than the average body. The *SANY* software calculates your RMR based on your individual body's mechanisms, not on the mechanisms of the "average" body.

The fact that people have different RMRs explains why you and your friend will probably lose weight at different rates even though you're following the same diet. Remember this when you use tables such as the one presented above. The values represent averages that may or may not accurately represent your unique metabolism.

ENERGY STORED IN YOUR BODY

You eat protein, carbohydrate and fat, and your body stores energy in the form of protein, carbohydrate and fat. The energy that you take in is either burned or stored. If your energy intake exceeds your total energy expenditure, the difference is stored in the body.

If the energy you take in is not enough to meet your energy expenditure, you are in negative Energy Balance. The reason negative

Energy Balance produces weight loss is that your body has to make up the energy difference somehow—that is, your body has to find the energy to burn.

When you are in negative Energy Balance, there is not enough energy in the food you have eaten to fuel your requirements, so your body turns to the energy it has stored and uses it for fuel. When this stored energy is burned, you lose weight.

Protein is usually stored in your muscles. Because your muscles are so important, your body rarely turns to stored protein for energy. It is only in periods of sustained severe negative Energy Balance that your body will break down protein and consume it for energy.

Your body can only store a small amount of carbohydrate. Carbohydrate is stored, in the form of glycogen, in your muscles and in your liver. Stored carbohydrate is readily used by the body and it is often the body's first choice when it is searching for energy to expend in physical activity. Your store of glycogen is quickly exhausted, and if you continue to be physically active once the glycogen is used up, your body will begin burning fat for fuel.

Most of the excess energy that comes into the body is stored as fat. We store fat all over our bodies. Some people—usually men—tend to store extra fat around the middle, while others—usually women—tend to store it in the hips and thighs.

Fat is a great way to store extra energy. It's very efficient because your body gets a lot of calories to burn for a little amount of weight. The problem comes when your body stores too much fat—a situation all too common in the United States today.

Negative Energy Balance
= Fat Burned as Fuel = Weight Loss

The good news is that fat is readily burned as fuel when you produce negative Energy Balance. In fact, producing negative Energy Balance—where you burn more calories than you take in through your diet—is the only way to get rid of stored fat.

ACHIEVING ENERGY BALANCE

Now that you understand the basic elements of your body's energy use and storage, your goal is to create a state of Energy Balance that allows you to lose the weight you want (negative Energy Balance), and then to create a state of Energy Balance that keeps you from gaining the weight back (zero Energy Balance).

Let's go back to our financial analogy. If you want to increase your checking account by $1,000, you need to deposit $1,000 more than you spend. You can choose how you want to make the money accumulate. You can get a second job and deposit more dollars, you can reduce your spending and keep the dollars in the account, or you can work overtime and spend less.

The point is that once you set your goal, you know what you have to do—and then you simply choose how to do it.

Let's continue with the financial analogy. You have total control over your bank account. You can decide to increase it or decrease it. The goals you set may be realistic or unrealistic.

For example, you may decide you want $1,000,000 in your account. If you make $50,000 per year and spend $40,000 per year, it is going to take you a very long time to reach your goal. This may not be a realistic goal, even though you really would like to have $1,000,000.

Once you see what it would take to achieve your goal, you might decide that it is, after all, not realistic and that a better goal would be to accumulate $50,000. You can do this in five years with your current spending patterns.

Alternatively, you might decide you want to accumulate your $50,000 more quickly. Again, you have a choice. You can try to put more money in or take less money out.

What's crucial is that you set your goal. Once you have done that, you will know **what** you have to do to reach your goal. Then you decide **how** to do it based on financial principles.

Managing your weight is very similar. You can set your weight loss goal and then, with an understanding of Energy Balance, determine what you have to do to reach that goal.

For example, let's say your goal is to lose 50 pounds. Is this goal reasonable and attainable? Without an understanding of Energy Balance you would be just guessing at the answer to that question. But now that you understand Energy Balance, answering the question becomes a simple logical exercise.

3,500 Calories = 1 Pound

Since you know that it takes 3,500 calories of negative Energy Balance to lose 1 pound, a simple calculation tells you it will take 175,000 calories of negative Energy Balance to lose 50 pounds.

If you create a negative Energy Balance of 500 calories each day, it will take you a year to reach your goal.

175,000 Calories/500 Calories per day = 350 Days

It will actually take a bit longer than one year to lose the 50 pounds, because your body will reduce the number of calories it burns as you lose weight. Your RMR will go down as your body gets smaller (but in the exchange, you will get a smaller, much healthier body).

WHAT IS REALISTIC?

If you create a negative Energy Balance of 1,000 calories each day, you can lose the 50 pounds in half the time. Creating a negative Energy Balance of 500 to 1,000 calories a day is realistic.

Perhaps you are impatient and want to lose your weight even faster. Let's think about it. Applying the concept of Energy Balance, we see that if you wanted to lose the 50 pounds in only 3 months, you would have to reduce your energy intake by about 1,900 calories each day.

For most of us, this would be highly unrealistic. The point here is that you can use your new understanding of Energy Balance to set a weight loss goal that is both acceptable to you and attainable.

The next step for you to take is to learn what it is you have to do to achieve that realistic, attainable goal.

QUANTITY OF FOOD AND DIET COMPOSITION

Many popular diet books convey the message that the types of foods you eat are more important in determining your weight than the total number of calories you eat.

Nothing could be further from the truth. We have an obesity problem in the United States because we are eating too much of every kind of food.

Too Much Food = Too Much Weight

While diet books imply that there is something magical about the composition or combination of foods they recommend, it is really all about calories. If you do not produce a negative Energy Balance of 3,500 calories, you will not lose a pound of weight—it doesn't matter what kind of food you eat. While what you eat will affect your general health, it will not affect the amount of weight you lose. A calorie is a calorie, no matter where it comes from.

A calorie is a calorie—regardless of where it comes from.

- Eat a nutritionally balanced diet that satisfies your hunger.
- Consult your doctor if you have special needs.

If you overeat by 500 calories, you will be in a positive Energy Balance regardless of whether those 500 calories came from fat, carbohydrate, or protein. There is no food that allows you to overeat without going into positive Energy Balance.

Scientific studies have shown that when people maintain a fixed negative Energy Balance of 500 calories per day (that is, their energy intake is 500 calories less than their energy expenditure), they will lose weight at the same rate—regardless of what it is they eat. This is true of high fat, high carbohydrate, and high protein diets.

When it comes to negative Energy Balance, your body does not care how you achieve it. Using our financial analogy, it is not important to your bank account how you get or where you spend your money—a dollar is a dollar whether you spend it on groceries, on jewelry, or on entertainment.

When it comes to your bank account, a dollar spent is a dollar spent. When it comes to your weight, a calorie of negative Energy Balance is a calorie of negative Energy Balance.

THE HUNGER FACTOR

This does not mean that diet composition is something you can ignore. There are several reasons why you need to know about diet composition.

For example, we know that different people will lose different amounts of weight on different diets. But why is that so? Some people find it easier to stick to a low-fat diet, while others find a low carbohydrate diet more comfortable for them. The reason different diets have different effects on weight loss is that they allow for different degrees of negative Energy Balance.

Let's take a couple of examples. Sara tried several different diets before she found that a low carbohydrate diet worked best for her. The low carbohydrate diet worked for her simply because Sara felt less hungry when she produced negative Energy Balance eating a low carbohydrate diet.

For Mary, it was just the opposite—Mary felt less hungry on a high carbohydrate diet. Because Sara and Mary found the diets that were best suited to their individual preferences, Sara lost more weight on the low carbohydrate diet and Mary lost more weight on the high carbohydrate diet.

There is nothing magical about diet composition—the simple truth is that different diets affect hunger differently in different people. This means that the best diet for weight loss for you is the diet that allows **you** to most easily maintain a negative Energy Balance.

It is the negative Energy Balance that produces weight loss. The type of diet chosen can affect the ease with which you produce a negative Energy Balance. Your task is to learn which type of diet works for you. We'll talk later about how the *SANY* software program can help you do just that.

DOES IT MATTER WHAT YOU EAT?

There are reasons to be concerned about what foods you eat. Some experts suggest that certain foods, or combinations of foods, may make you want to eat again after a period of time has elapsed. Take, for example, the suggestion that, if you eat foods that cause your blood sugar to rise rapidly, you will be hungry a short time later and will eat again; as a result, you will end up consuming more calories than if you had not eaten those foods in the first place.

Not all experts agree that this is true. It may be that these foods, and others, simply affect different people in different ways.

However, we all know that what you eat does affect your health. For example, eating a high fat diet—especially if it is high in saturated fats—can lead to heart disease.

It all boils down to this: from an overall health viewpoint, it is true that you should consider what type of food you eat. But as far as achieving your desired state of Energy Balance is concerned, a calorie is a calorie is a calorie.

From an Energy Balance point of view, it is important that you give consideration to certain issues related to eating.

Portion size, for instance, is important. As we saw in Chapter 2, too many Americans are super-sizing their meals. The more food you have on your plate, the more calories you are likely to ingest. Few people stop to think about the extra calories they are getting when they choose the super-size meal.

Once you start thinking in Energy Balance terms, you will think twice about portion size.

DON'T FORGET ABOUT PHYSICAL ACTIVITY!

We have considered how the composition of your diet affects your Energy Balance. Now let's consider whether the type of **physical activity** you perform affects Energy Balance.

It doesn't. Here again, it is the total number of calories that is important. You can burn 100 calories by walking a mile or by lifting weights for 15–20 minutes.

---

100 Calories Burned by Walking
= 100 Calories Burned by Lifting Weights

Both exercises have exactly the same effect on energy expenditure. In other words, their contribution to your Energy Balance is exactly the same. In Energy Balance, the key point is how many calories you burn in total during your physical activity. The type of activity you do is not nearly as important.

If your goal is to produce negative Energy Balance of 500 calories per day and you decide to reduce intake by 300 calories per day and increase physical activity by 200 calories per day, you can do the latter by walking, running, swimming, weight-lifting, or any number of other activities. It does not matter which activity you choose. It's completely up to you—so long as your physical activity burns 200 calories each day.

Time may be an important factor. You can burn 100 calories either by walking a mile or by running a mile. However, you burn these calories much more quickly if you run the mile.

Well, here you are. You've read this far and you've learned all about Energy Balance. Now you're ready to move to the next step.

It's time for you to identify your weight loss goal and establish your timeline for reaching it. You need to translate your weight loss goal into an Energy Balance goal. Then decide how you want to modify your patterns of eating and engaging in physical activity to achieve your Energy Balance goal.

We'll guide you through the process in the next chapter.

CHAPTER 5. LOSING WEIGHT

James O. Hill, Ph.D.

Losing weight sounds as if it should be easy—anyone should be able to do it. In Energy Balance terms, all you have to do to lose weight is to eat a little less food than you would ordinarily eat. What's the big deal about leaving a small amount of food uneaten at each meal? Losing one pound per week is as simple as producing 500 calories of negative Energy Balance each day for one week.

The strategy for losing weight is simple and it is all about Energy Balance, but that does not necessarily mean it is easy. The reality is that acting on almost any new strategy is a challenge, and implementing a strategy that conflicts with your instincts is especially hard. Restricting the amount of food we eat is a big deal. It is a big deal because our physiology was developed to make it easy for us to eat, not to make it easy for us to restrict our food.

Every diet plan ever developed was developed for one reason—to help you produce negative Energy Balance, primarily by reducing the number of calories you take in. Every dieter who has ever lost weight has done so through achieving a negative Energy Balance.

Quite simply, there is no other way to lose weight. **While some popular diet plans may suggest that there is something special about their program that allows you to lose weight without restricting your energy intake, this cannot be true.** The only way to lose weight is with negative Energy Balance and the only way to produce negative Energy Balance is to take in fewer calories than you burn. Unless a diet plan advocates exceptional amounts of physical activity, weight management relies primarily on restricting caloric intake.

Let's make one thing perfectly clear—losing weight is not easy. It can be done; many people succeed, but they succeed only with conscious effort. If you start a weight loss program but are not committed to it, you are likely to fail.

But if you are committed, all you need to get started is a good plan. Just like building a house or going on vacation, you are much more likely to succeed if you have a reasonable plan in place. The same is true of weight loss.

People who succeed in losing weight and keeping it off are those who take control of their Energy Balance. They refuse to be at the mercy of an environment that encourages eating too much and not getting enough physical activity.

Your instincts may tell you to eat whenever food is available and rest whenever you can. But your brain can overrule your instincts and allow you to decide how much to eat and when to be physically active.

We can help you make the plan, but you have to supply the dedication.

HOW MANY PEOPLE SUCCEED IN LOSING WEIGHT?

As you think about losing weight, you probably wonder about your chances for success. It is very common for the media (television, newspaper, magazines) to imply that almost no one succeeds in long-term weight loss. We often hear, for example, that over 95% of people who lose weight gain it back. Is there any hope for success?

The answer to that question depends on our definition of success. Not very many people who are classified as obese (those with a BMI equal to or greater than 30) succeed in getting their BMI down to 25 or below and keeping it there. If this is the measure of success, then most people do fail.

However, this is not the only—or even the best—measure of success. If your goals are to lose weight, improve your health, and look and feel better, you can absolutely achieve those goals without getting to a BMI of 25 or less.

Rather than focusing on BMI as your primary measure of success, an alternative approach might be to decide that your weight loss goal will be a percentage of your current weight.

For instance, if you define successful weight loss as losing 10% of what you now weigh (if you now weigh 200 pounds, that works out to just 20 pounds) and keeping that weight off for at least one year, then you are greatly increasing the probability that you will succeed in your weight loss program.

10% of Your Current Weight
= A Realistic Weight Loss Goal

In this chapter, we will talk about losing weight, but remember that this is only half of the task. Success involves both losing the weight and keeping it off. You should be prepared to transition from a weight loss program to a lifestyle that allows you to balance the energy you take in with the energy your body burns. The diet you pick now can help you make that transition.

But first things first. Before you can keep weight off, you have to take it off. What you need to do now is develop your plan to create the negative Energy Balance that will take the weight off.

BARRIERS TO WEIGHT MANAGEMENT SUCCESS

As you get ready to create your weight loss plan, it might be useful to learn from those who didn't succeed. What did they do wrong that prevented their success? We found that there are five common barriers to long-term weight management success.

A big reason that many people fail is that they simply have not prepared themselves for the effort it takes to change their lifestyle. Not everyone who needs to lose weight is ready to make those changes. Until you are ready to change or modify some of the behaviors contributing to your weight gain, you are not likely to be successful in losing that weight and keeping it off.

A second reason people fail at weight management is that their whole focus is on taking the weight off (short-term focus) and not on keeping it off (long-term focus).

Five common barriers to weight loss success:

- Being unwilling or unable to change lifestyle.
- Focusing only on short-term weight loss.
- Having unrealistic expectations.
- Focusing only on diet and not on a combination of diet and physical activity.
- Relying on a "magic bullet."

Celebrate your success when you lose weight—but keep in mind that maintaining your new weight will require a lifestyle change. Unless you change your lifestyle, you will find yourself on a never-ending series of diets.

The majority of diet plans focus almost exclusively on changing the foods we eat. It's true that changing your food intake will be a very important part of your successful weight loss program—but it is not the only component. As we saw in Chapter 1, it is difficult to lose weight just by increasing your physical activity—but physical activity will be an important part of your successful weight loss program.

People who have succeeded in weight management over the long-term know that both diet and physical activity are important. When you enter the weight maintenance phase, physical activity will become even more important than it was during the weight loss phase.

If you do not increase your physical activity during weight loss, when you reach your weight goal and want to switch to a weight maintenance program, you will almost certainly discover that you will not be able to increase your activity quickly enough to avoid gaining the weight back. It is very hard to change your Energy Balance by changing only one side of the Energy Balance equation.

Less Food + More Physical Activity
➔ Adjusted Energy Balance

UNREALISTIC EXPECTATIONS

Many people start a weight loss program with a very unrealistic expectation about how much weight they will lose. If your weight loss goal is unrealistically high, you are setting yourself up to fail. It is extremely important that you **set a realistic goal**. We will show you how to do that.

Keep in mind that once you accomplish your first weight goal, you will be able to set another goal. If you give yourself a series of achievable goals, you are more likely to succeed than if you set an initial goal that is too difficult to reach.

Many diet plans promise you a "magic bullet." If you follow their plan, they promise, the weight will melt away and stay off forever. Most of us have tried the easy plans, but they haven't worked. And the next one to come along is not likely to work any better than the last three did. Sure, it would be wonderful to find a "magic bullet"—but there is just no such thing.

Weight management is all about adjusting your Energy Balance by changing your diet and lifestyle; it's not about finding a miracle plan.

There are no magic bullets!

The question to ask yourself now is this—are you willing to do what it takes to be successful? If you decide you are ready, we're here to help.

You **can** be successful, by developing a realistic weight management plan that is based on the simple concept of Energy Balance.

WHO NEEDS TO LOSE WEIGHT?

Most adults in the United States need to lose some weight. We told you in Chapter 1 that 65% of Americans are overweight and that being overweight endangers their health.

The National Institutes of Health advises those of us whose BMI is over 30 to lose weight to protect our health. About 35% of us fall into that range. If your BMI is 30 or more, your excess body weight puts you at risk of very serious diseases such as diabetes, heart disease and some cancers. If you already have one of those diseases, losing weight can help you manage it. If you do not have one of those diseases, losing weight will significantly lower your risk of getting one.

If your BMI is now between 25 and 30, you are classified as overweight. This means that your risk of serious disease is greater than average, but it's not nearly as high as for people who are obese. If you are overweight, your first priority is to make sure you don't gain any more weight. You want to make sure you don't go from being overweight to being obese.

Remember, most American adults gain one to three pounds each year. At that rate, it does not take very long to go from a BMI over 25 to a BMI over 30. Gaining that extra weight would dramatically increase your risk of developing a weight-related disease.

If you are overweight but not obese, you can realistically lose enough weight to return to the healthy weight category. That is, a weight goal that would get your BMI to under 25 is an achievable weight goal for you. You can look forward to success with such a goal—and to a significant improvement in your health.

If you are classified as overweight and already have a weight-related condition—such as diabetes, high blood pressure, elevated cholesterol, or insulin resistance—for you, weight loss becomes an even higher priority. Losing weight can significantly improve these conditions and may slow the progression toward even more serious diseases such as heart disease.

TWO LANDMARK STUDIES

Two landmark research studies are relevant here. The first is the Diabetes Prevention Program (DPP), which was described in Chapter 3. Remember, DPP was a research study to determine whether lifestyle modification (modest weight loss and a modest increase in physical activity) could prevent development of type 2 diabetes in individuals who were at very high risk for this disease.

The researchers found that lifestyle modification had a dramatic effect on preventing type 2 diabetes. The study demonstrated that individuals who were overweight and at increased risk for diabetes received tremendous benefit from the modest lifestyle change that formed part of the study.

A second landmark study currently underway is building on the results of the DPP study. It's called the Look AHEAD study[14]. The aim of this second study is to see whether lifestyle modification can slow the progression of heart disease in overweight or obese people who already have type 2 diabetes. Type 2 diabetes and heart disease are closely related—most people who develop type 2 diabetes also develop heart disease. This study is also sponsored by the National Institutes of Health and will involve about 6,000 subjects.

Mounting evidence from these and similar studies demonstrates the powerful effects that modest lifestyle changes can have on the development and progression of chronic diseases such as diabetes and heart disease.

If you already have a healthy BMI, try not to gain weight! We'll say it again—if your BMI is between 18.5 and 25, your primary goal is to not gain any weight.

---

[14] NDDKD. http://www.lookaheadstudy.org

SETTING REALISTIC WEIGHT MANAGEMENT GOALS

It's time to begin creating your individualized weight loss plan. Your first task is to set a realistic weight management goal for yourself. Remember, successful weight management begins with setting a realistic goal.

There is no formula for setting your goal. The figure below shows some options for setting your first weight management goal.

We already know that most Americans gain weight every year; the diagonal arrow shows that basic, unhealthy trend. That trend will continue—that is, people will continue a slow, steady weight gain—unless they make changes to their lifestyle.

If you are one of the numerous Americans who gain one to three pounds each year, weight maintenance may be a good first goal for you. In other words, your first goal could simply be to **stop gaining weight**. It's an achievable goal, and it could have a profound positive effect on your health. Line 1 in the figure shows successful weight maintenance.

Managing Obesity

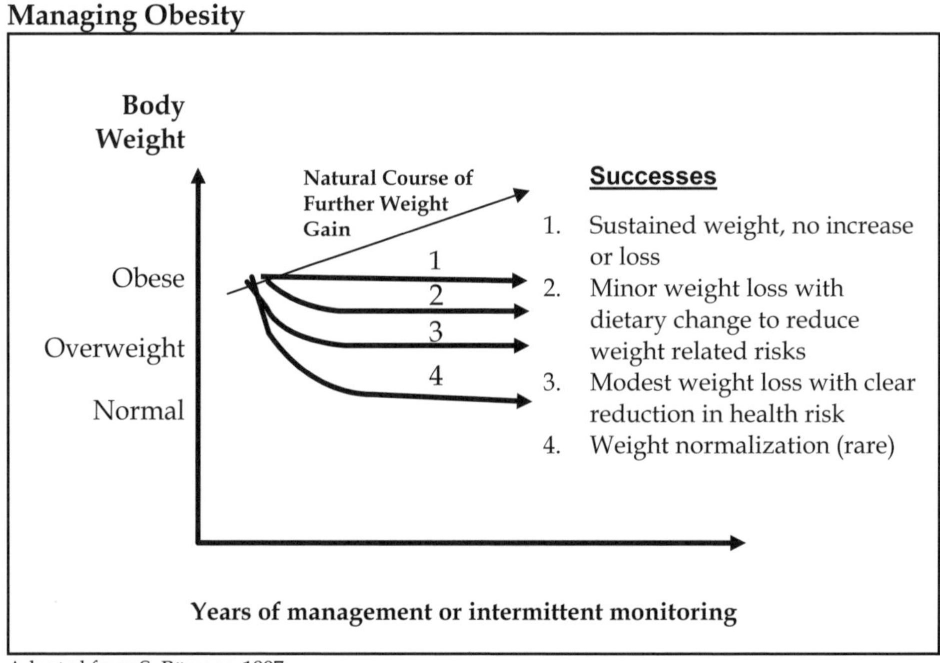

Adapted from S. Rössner, 1997

But you may decide that just stopping weight gain is not enough. Perhaps you want to lose weight. Beware—many people set unrealistically high weight loss goals. When researchers asked people entering a behavioral weight loss program how much weight they expected to lose, many said they expected to lose as much as 30% to 40% of their current body weight. In fact, the average weight loss actually produced in the program was 8% to 9% of subjects' initial weight.

If you expect to lose 30% of your weight and you lose 8%, you might see this as a great failure. On the other hand, your doctor might see your 8% weight loss as a great success—because that amount of weight loss can significantly improve any weight-related conditions you already have and lower your risk of developing other weight-related diseases.

It's a fact that very few people lose 30% to 40% of their body weight through lifestyle modification programs. The very best weight loss programs available generally produce about 8% to 12% weight loss. This means that if you weigh 200 pounds, you can realistically expect your weight loss program to produce a weight loss of 16 to 24 pounds—not 60 to 80 pounds. So...set your first weight loss goal at a realistic level, say 8% to 12% of your initial body weight, and record this in your weight management record on page 165.

Realistic weight management goals:

- Aim for a healthier weight, not an ideal weight.
- Accept that your progress will be slow. (And remember that slow is real.)
- A good first (short-term) goal is a weight loss of 5% to 10% of your starting weight, or 1 to 2 pounds each week.
- A good interim (medium-term) goal is weight maintenance—keeping the weight off.
- Your long-term goal is to keep the weight off forever, and perhaps to lose additional weight.

The point is that, if you set an unrealistically high weight loss goal, you are setting yourself up for failure. Sure, many of us would be thrilled to lose a lot of weight quickly—and wouldn't we look great if we did! But when you look at the facts, you see that setting a weight loss goal that's too high is not such a good strategy. It makes far more sense to set a smaller, achievable goal and then, once you reach it successfully, set your next goal.

That way, you avoid feeling disheartened because you're unable to meet an unrealistically aggressive goal. Instead of setting yourself up for disappointment, set a realistic goal so that you can appreciate and celebrate your successful achievement—a weight loss that significantly improves your health and quality of life.

10% of Starting Weight
= A Good Initial Weight Loss Goal

It is important that you give considerable thought to your weight loss goal before you start your weight loss plan. Don't skip this step!

We recommend aiming for a weight loss goal that is both satisfying to you and achievable. The National Institutes of Health's *Guidelines for Managing Overweight and Obesity* suggests that an initial weight loss goal should be about 10% of starting weight. Our research supports that recommendation. Research repeatedly shows that a 10% weight loss (sometimes even a 5% weight loss) can dramatically lower the risk of disease development or progression.

HOW QUICKLY SHOULD YOU LOSE WEIGHT?

Never mind that it took you 10 years to gain the extra weight you're carrying; now you want to lose it all in 10 weeks. Most people feel that way—they want to lose the weight fast. But experience shows that weight lost more slowly is often easier to keep off. This is in part because slower weight loss more often than not results from real, sustainable lifestyle change rather than from unsustainable "magic bullet" dieting.

Once you have set your initial weight loss goal, the next step is to decide on a reasonable timetable for achieving it. Again, there is no simple formula to determine how quickly you should lose weight.

How fast?

- National Institutes of Health recommends weight loss of 1 to 2 pounds each week.
- This means a negative Energy Balance of 500 to 1,000 calories each day.
- Remember that your metabolism will slow down as you lose weight. Allow for that in your timetable.
- Rapid weight loss increases the likelihood of gaining the weight back.

The National Institutes of Health *Guidelines for Managing Overweight and Obesity* suggests that slow, incremental weight loss is most likely to lead to long-term success in keeping the weight off. These guidelines suggest a weight loss of one to two pounds each week.

Many programs target this realistic rate of weight loss, but others aim for faster weight loss. Diets based on intake of very few calories—often fewer than 800 calories a day—can lead to much faster weight loss. However, The National Institutes of Health *Guidelines* conclude that, while those diets are not necessarily unsafe, the weight lost is often regained.

To be fair, there is no conclusive research that shows a "best" rate for losing weight. But the evidence points to one conclusion—expecting to lose one to two pounds each week, as suggested by the National Institutes of Health, is a reasonable timeline to start with. Space is provided in your weight management log (page 165) to record your initial goals.

TRANSLATING YOUR GOALS INTO ENERGY BALANCE

You are now ready to translate your weight management goal into Energy Balance terms. In order to lose one pound of body weight, you must take in 3,500 fewer calories than your body burns. In other words, you must create 3,500 calories of negative Energy Balance for each pound of weight you want to lose. If you want to lose 20 pounds you must accumulate a negative Energy Balance over time of 70,000 calories.

Calories count!

- To lose weight, reduce your energy intake.
- 3,500 calories = 1 pound of body weight.
- To lose 1 pound each week, decrease calories eaten by 500 each day.
- To lose 2 pounds each week, decrease calories eaten by 1,000 each day.

To find out how much you have to change your Energy Balance, divide the total number of pounds you want to lose by the total weeks in your weight loss timeline. Let's look at some achievable goals—

- To lose 20 pounds in 20 weeks, you need to lose one pound each week. This means you need a negative Energy Balance of 3,500 calories a week, or 500 calories each day.
- To lose 20 pounds in 40 weeks, you need to lose just half a pound each week. This is a negative Energy Balance of 1,750 calories a week, or 250 calories a day.

Both of these are very reasonable goals. Achieving a negative Energy Balance of 250 or 500 calories a day is a realistic expectation.

What Energy Balance is right for YOU?

Alternatively, setting a weight loss goal of 20 pounds in 4 weeks would require 5 pounds of weight loss each week. That would require a negative Energy Balance of 17,500 calories per week or 2,500 calories each day. A negative Energy Balance that large is clearly not reasonable—you'd be eating almost nothing for four weeks.

Translating your weight loss goal into Energy Balance terms can help you gauge whether your goal is realistic. Most people eat somewhere between 1,500 and 3,000 calories each day. Larger people consume more calories.

Try translating your weight goal into Energy Balance terms. What size of negative Energy Balance are you aiming for each day? As you look at a couple of options, you will begin to see what amount of negative Energy Balance may work for you.

Most people can achieve a negative Energy Balance of up to 1,000 calories a day. If your weight loss goal requires a greater negative Energy Balance, have another look at your weight loss goal and at your timetable for achieving that goal.

AN UNEXPECTED PROBLEM—YOUR METABOLISM

Here is another important reason to set realistic weight loss goals—your body's metabolism actually works against you when you lose weight.

You may remember our discussion of metabolic changes in Chapter 4. As you lose weight, you become smaller. Remember that your total energy expenditure (your total metabolism, or the total number of calories your body burns) is the sum of your resting metabolic rate (RMR), your Thermic Effect of Food (the percentage of total energy expenditure used in the digestion of food), and the energy your body burns in physical activity.

Your Total Energy Expenditure =
Your RMR + Thermic Effect of Food
+ Calories Burned in Physical Activity

Your RMR is directly related to your body size. As you become smaller, your RMR goes down. When you eat less food, your Thermic Effect of Food also goes down. Additionally, once you have become smaller, your physical activity burns less energy, further contributing to the reduction in your total energy expenditure.

What does all this mean? It boils down to this—to keep the same degree of negative Energy Balance as you lose weight (and to keep your rate of weight loss constant), you need to take in even less energy.

Effect of decreasing RMR

Description	Energy intake	Energy expenditure
Before weight loss	2,500 calories	2,500 calories
During weight loss	2,000 calories	2,500 calories
Daily negative Energy Balance		500 calories per day
Weekly result before RMR decreases		Negative Energy Balance of 3,500 calories per week
		As RMR decreases, the energy cost of physical activity decreases to less than 2,500 calories
Weekly net result		Less than 3,500 calories negative Energy Balance, depending on weight lost

Thus, total energy expenditure progressively decreases from 2,500 calories to an amount that may be 5% to 20% less. The amount of decrease in energy expenditure depends entirely on your individual physiology and activity levels.

So, even though your energy intake stays the same over time, your negative energy expenditure will decrease each week. The result? Each week, you will lose less weight than you did the week before. It hardly seems fair, does it?

Is there a way to compensate for this decreasing energy expenditure? What can you do to keep "spending" those 3,500 calories—and losing that pound—week after week during your program?

Lower your energy intake to compensate for decreasing RMR.

To keep the rate of weight loss the same, your energy intake must decrease in some proportion to the loss of body weight. Unfortunately, that proportion is very difficult to calculate manually. The *SANY* software performs these calculations for you. If you are not using the software, the important lesson here is that you must stay aware of the effect of your decreasing RMR as you lose weight—and you must make some kind of adjustment to compensate for that effect.

REVISING YOUR WEIGHT LOSS PLAN

Once you have arrived at your weight loss plan, you may find that, for one reason or another, you'd like to change it. That's okay! It is perfectly fine to revise your weight loss plan as you begin to change your lifestyle.

If you were building a house, you might find that you wanted to change the plans as construction progressed. You might, for example, find that you have less money to spend than you had anticipated and decide to take out some of the features that weren't essential. Or you might find that you needed extra closets, or that the bedroom should be larger. So you revise it. The important thing is that you still have a plan.

Let's assume your weight management goal is to lose 20 pounds in 10 weeks. That means losing two pounds each week, which requires a negative Energy Balance of 1,000 calories per day.

If you find that reducing your food intake by 1,000 calories is too difficult, you might modify your timeline, so that your revised plan is to lose 20 pounds in 20 weeks. That requires a negative Energy Balance of only 500 calories each day.

You've adjusted your plan to give yourself a longer period to reach your goal. You've realized the first goal you set did not accommodate your needs very well and was therefore not achievable, and you've adjusted it accordingly. One of the wonderful things about the concept of Energy Balance is that it makes it easy to adjust your plan if you need to.

ACHIEVING YOUR PLANNED STATE OF ENERGY BALANCE

Your next step is to make the lifestyle changes you need to adhere to your weight management plan. Remember that a successful weight management plan almost always requires some changes in lifestyle.

Up to this point, you've dealt with the dietary, or energy input, side of the equation. You've established your weight management goal, translated your goal into Energy Balance terms, and determined your timetable. Now you need to achieve the state of Energy Balance that will achieve your goal.

You already know that there are three ways to change your Energy Balance. If your plan requires 500 calories of negative Energy Balance a day, you can decide to:

- eat 500 fewer calories each day,
- increase your physical activity by 500 calories a day, or
- choose any combination of eating less and moving more that adds up to 500 calories.

It is up to you. What's your strategy?

You might pick up some ideas from people who have reported their weight loss successes to the National Weight Control Registry. Here's what their stories tell us: less than 10% of them lost weight by just changing their diets (energy intake), and almost none of them lost weight just by increasing their physical activity (energy expenditure). While you could decide to use a single strategy, the evidence strongly suggests that you are more likely to meet your weight management goal if you modify both your diet and your physical activity.

Why is it better to modify both behaviors than to modify only one? The answer is pretty simple when it comes to physical activity. It is difficult to create a large negative Energy Balance by increasing physical activity alone. While eating 500 calories less in a day is relatively easy, performing an additional 500 calories of physical activity every day is unrealistic for many people.

Increasing physical activity is a good way to lose a few pounds—but if your weight loss goal is substantial, you are not likely to achieve it solely by increasing your physical activity.

Choosing how to achieve a negative Energy Balance:

- Your negative Energy Balance goal (usually 500–1,000 calories) can be most easily achieved through a combination of eating less and exercising more.
- It's possible to accomplish your goal through diet alone, but NWCR suggests that is not the easiest way.
- It's difficult to accomplish your goal through exercise alone.
- Your best plan combines diet and exercise!

It is very interesting that so few people in the National Weight Control Registry lost their weight through diet alone. There are several reasons that might explain this fact.

First, as previously explained, our physiology is not designed for food restriction. Sustaining food restriction for long periods of time is extremely hard for most people. Increasing physical activity relieves this quandary somewhat by shifting some of the emphasis away from food restriction. Many people find that it is easier to sustain modest increases in activity than it is to maintain severe restrictions of their food intake.

Choose a combination of diet and activity that parallels your current lifestyle as closely as possible. The idea is to minimize the amount of change that's necessary to accomplish your weight loss goal.

A second reason for combining strategies is that using two strategies instead of only one gives you more flexibility. If you rely on food restriction alone, you either restrict your food as required or you don't. But if you make adjustments to both sides of the Energy Balance equation, you can vary your eating and your activity behaviors to better fit your lifestyle. That allows you to rely on one behavior more on some days—that is, eat less on some days—and on the other behavior—that is, exercise more—on other days.

Finally, increasing physical activity during weight loss may be critical to your long-term success, because it prepares you to maintain your weight loss. We will show you in the next chapter why increasing physical activity is critical for keeping your weight off.

COMMON COMPONENTS OF OBESITY TREATMENTS

Look at the table below. It shows the most common components of weight management plans. Weight management involves changing your state of Energy Balance and you do this through the first two components—changing your energy intake, your physical activity, or both. The last three components listed will help you make these changes.

Components of weight management plans:

- Modify your diet.
- Increase your physical activity level.
- Modify your behaviors.
- Treat obesity with medication.
- Treat obesity with surgery.

We will not discuss obesity surgery or obesity medications at any length. Obesity surgery may be an option for you if your BMI is greater than 40 (see page 57 to calculate your BMI if you have not already done so). If you fall into this classification, surgery is an option you can discuss with your physician. Obesity surgeries such as the stomach bypass involve modification of the gastrointestinal tract. The procedures create negative Energy Balance by making less of your ingested energy available for absorption by your body.

Two prescription medications have been approved by the Food and Drug Administration (FDA) for weight management—Meridia and Xenical. These medications are indicated for people with a BMI equal to or greater than 30, or for people whose BMI is greater than 27 if they have weight-related conditions such as high blood pressure or diabetes. Most physicians recommend trying to lose weight through lifestyle modification before trying one of these medications.

If you're reading this book, most likely you are thinking about changing your energy intake or your patterns of physical activity. In that case, it's time now for you to choose your diet plan and your physical activity plan. Later on, we'll introduce some behavior modification tips to help you stay with your plan.

CHOOSING A DIET PLAN

There are so many diet plans out there! How do you find the one that's right for you?

Unfortunately, we don't have an easy answer to that question. Any diet that helps you reduce your energy intake can help you produce a negative Energy Balance and lose weight—if you can stick with it.

You can go to your local bookstore and find hundreds of different diets for losing weight. All of these diets are likely to work—if you follow them.

But remember, not many of those diets are likely to help you keep the weight off once you've lost it. The diet you choose to lose the weight may not be the one that will help you keep the weight off.

The *SANY* software is designed to work with this book and the Energy Balance concept—but it also works well with any popular diet plan. Remember, the key is to find a plan that fits your likes and dislikes as closely as possible—one that will make your transition from weight loss to weight maintenance as smooth as possible.

Part II looks at some of the limitations experienced by dieters and explains how the *SANY* software will assist you in personalizing your weight loss and weight maintenance programs. The *SANY* software allows you to custom-fit your weight program to suit your personal preferences.

What is the best diet?

- Many experts promote diets based on the Food Guide Pyramid (see page 48).
- There is not much hard data on diet and weight loss.
- Current findings suggest that the most important factor is how many calories you eat.
- Energy density (the amount of energy in a given weight of food) is important.
- The use of meal replacement products is gaining support among weight loss experts.

BALANCED DIETS

Many reputable organizations such as the National Institutes of Health, the American Heart Association, and the American Dietetic Association recommend that you look for a diet that produces a weight loss of one to two pounds a week and that provides you with a good mix of nutrients.

A typical balanced diet provides:

- 30% of calories or less from fat, with less than 10% from saturated fat,
- 15% to 20% of calories from protein, and
- the rest of your calories from complex carbohydrates.

These health organizations recommend that the diet contain adequate amounts of vitamins and minerals. In others words, these are very healthy diets. It is recommended that you stay with a dietary balance like the one described above, but when you want to achieve negative Energy Balance, you should eat less. In other words, ensure that you have the proportions of calories from fat, protein, and complex carbohydrates as recommended above, but restrict the overall intake.

These are diets you can eat forever. If you are successful in losing weight on this type of diet, you will likely not have to change the types of food you eat when you are switching from weight loss to weight maintenance. And after you have reached your weight goal and are no longer seeking a negative Energy Balance, you will probably be able to slightly increase the total amount of food you eat.

POPULAR DIETS

Many popular diets take a very different approach. Often they ask you to dramatically change your eating habits. Nutrition experts usually consider these kinds of diets to be "unbalanced."

For instance, you might be asked to eat a diet that is very low in fat, very low in carbohydrate, or very high in protein. You might even be asked to eat a diet consisting mainly of soup, beer, or ice cream.

These diets work in the short-term—not because there is something special about the kinds of food you are eating, but because the diets focus your attention on what you are eating and help you restrict the total calories you consume. Any time you eat fewer calories, you are reducing your energy intake, creating a negative Energy Balance—and losing weight.

Should you avoid these diets? It depends on your approach to weight management. If you use these diets temporarily, to produce a negative Energy Balance and lose weight, they can work. However, if you plan to stay on one of these diets for the rest of your life, you might be disappointed in the results. You could even suffer serious health consequences as a result of the long-term nutritional imbalance.

Why are reduced-fat diets associated with weight loss?

- Eating less fat results in a lower energy intake—which leads to weight loss.
- The dieter might eat less food on these diets because the food just doesn't taste good.
- The dieter may be eating a constant weight of food and ensuring that most or all of the food eaten has low energy density—resulting in a low energy intake (see page 108).
- The dieter may feel satisfied by fewer calories.

Popular diets do work in helping some people lose weight. Interestingly, for each of these popular diets, there are groups of people who seem to do very well and other groups who seem not to do so well.

Why would this be the case? It probably is about managing hunger. Some people cannot stay on diets because their total food intake is so restricted—they simply get hungry when they eat less than normal.

The people who lose more weight on diets are the people who follow them more closely for longer periods of time. It appears that some people are better able to manage their hunger when they eat a diet containing certain types of foods, while other people can better manage their hunger on a diet containing different kinds of foods.

Some people, for example, find they get hungrier on a diet low in carbohydrate, while others find just the opposite. This is why there are so many different diets—no single diet plan is right for everyone.

If you want to go on a popular diet, be prepared for some trial and error. You may need to try several diets before you find one that works for you.

How to Calculate Energy Density

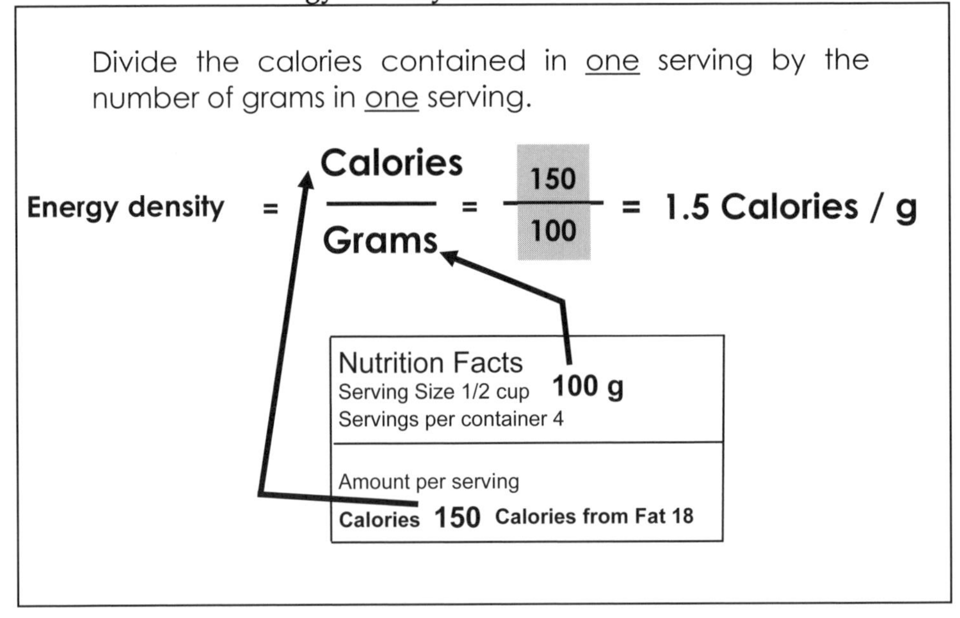

STRUCTURED MEAL PLANS

A number of scientific studies have shown that you can lose weight using meal replacements, usually two per day. Many people enjoy the ease of substituting a ready-made meal, perhaps in the form of a liquid shake or a food bar, for a traditional meal.

This approach has several attractive features:

- Because the package tells them the exact number of calories and nutritional values per serving, they don't need to look up the caloric values of every ingredient in their meal. That makes record-keeping much more exact and less time-consuming compared to a traditional meal.
- Meal replacements are very convenient. Dieters don't need to cook and there is little or no clean-up.

EATING A LITTLE LESS

Another strategy that is worth considering is to keep eating the same foods as you normally do, but to eat just a little less at each meal. This plan has the advantage of being something you can maintain forever—but, for some people, it may not provide enough structure.

MAKING A CHOICE

It is time for you to make your choice. Any of the popular diet plans will help you lose weight if you can stick with them. Select the diet plan that you think will best fit your lifestyle. You may have to try more than one before you find the one that is right for you. Different people will fare better on different diets and, unfortunately, no one has discovered how to match people to the best diet for them.

The information in this book, combined with your daily food and activity logs (see Part II for examples), will provide you with the feedback you need to assess your progress and make necessary adjustments. Part II will also explain the advantages of using the *SANY* software to get extremely accurate feedback. *SANY* will even allow you to simulate dietary outcomes—this may help you determine which diet is best for you.

Choose a diet that's right for YOU.

This new breakthrough technology actually learns about your unique metabolism—it tells you how **your** body reacts to what you eat and to the exercise you get. It lets you decide how to adjust your diet and activities so that they really fit your likes and dislikes. This new technology gives you all the personalized information you need to reach your weight management goal.

CHOOSING A PHYSICAL ACTIVITY PLAN

By now, you've set your weight management goal, decided which strategies you'll use to maintain a negative Energy Balance, chosen a possible diet, and calculated your timeline. Your weight management plan probably includes increasing your physical activity and, hopefully, a specific goal for how many calories you want to burn each day.

It is important that you begin gradually. If you currently get very little physical activity, don't plan to start with an hour of aerobics every day. You need to start slowly and gradually increase your activity over time.

There are an infinite number of ways you can increase your physical activity. You can join a health club or gym, you can buy exercise equipment and exercise at home, you can join a sports team, you can take up tennis, or you can just start walking more. The choices are endless.

Be sure to choose a physical activity that you will do. If you don't do it regularly, you will probably not burn the necessary calories to meet your objectives. It doesn't matter what activity you choose, as long as it is one you can incorporate into your lifestyle. Just as all diets can help you reduce total caloric intake, all forms of physical activity can help you burn more calories—especially if you can make them a regular part of your lifestyle.

Any physical activity that you enjoy will do.

Physical activity plans and diet plans have a common problem. People find it relatively easy to stick with these plans for a short time, but they find it harder to do so over the long-term. Even if you do not like to jog, for example, you might be able to force yourself to jog each day for a few weeks or a few months—but if you don't enjoy jogging, you are not likely to keep jogging daily for the rest of your life.

Why not choose an activity you like? Then you can stay with it forever. You might choose walking, tennis, or golf.

BURN MORE CALORIES

We have provided a list in Part II (Activity List/Metabolic Rate starting on page 154) to help you calculate the amount of energy you burn performing different kinds of physical activities. Its use is more fully explained in Part II, but note that it is different from the traditional weight-related table illustrated in Chapter 4. The values in the Activity List are not weight-related and do not give an estimate of calories burned; instead, they provide an intensity factor that is used in conjunction with your RMR to more accurately calculate your energy expenditure. The list provides you with lots of options; just choose at least one activity that you think you would like and that you think you can stick with for the long -term.

You know how many calories you want to burn each day through physical activity. Using this list in conjunction with your RMR, you can calculate energy expenditure quite accurately.

Here are three strategies that you might consider to help you burn more calories through physical activity.

Strategies for burning more calories...

- Set aside some time each day for a planned physical activity.
- Reduce your use of labor-saving devices—for instance, take the stairs instead of the elevator.
- Reduce the time you spend in sedentary activities.

We've talked about the first strategy. Choose the activity or activities that you think you would enjoy and do these every day—or at least on a regular basis.

The second strategy is one you might not have considered before. You can greatly increase the number of calories you burn during the day by incorporating additional walking as a normal part of your day. You might, for example, take the stairs at work rather than the elevator or park your car further away from the office or stores.

Colorado has initiated a statewide program called Colorado on the Move™. Its goal is to get people to increase what is referred to as "lifestyle physical activity"—that is, walking. Lifestyle physical activity is measured using inexpensive electronic step counters. You can learn about this strategy, buy a step counter, and get some fun tips for walking more by visiting their website at www.coloradoonthemove.org. Appendix B contains a full description of Colorado on the Move™.

A third way to burn more calories is to spend less time being inactive. Look for ways to make slight reductions in the time you spend sitting. There are some simple ways to do this. You can, for example, watch one less television show in the evening. You can get up from your computer every 30 minutes and walk down the hall. Anything you do that decreases the time you spend sitting will help you burn calories faster.

It is important to understand that your energy use is a function of all your activities. If you find you have to spend hours behind a desk, try tossing an object from hand to hand for a few minutes to increase your energy expenditure. If you have to participate in numerous meetings, you might consider walking while you talk. There are a thousand little things you can do to increase your level of activity that don't make unrealistic demands on you to change your daily schedule--yet, when added together, burn significant amounts of calories.

BEHAVIOR MODIFICATION

Now you have created your diet plan and a physical activity plan. You have also been developing goals for changing your behaviors.

Let's review some behavior modification skills. You may find that adopting some of them will help you make the changes you want to make.

Behavior modification skills...

- Goal setting
- Self-monitoring
- Stimulus control
- Stress management
- Problem solving
- Cognitive restructuring
- Social support

GOAL SETTING

We have already talked about goal setting at some length. It is important to keep in mind that people who set goals for achieving a task usually do better than those who do not set goals.

You should now have a well-developed plan for your weight management. Writing your plan down will improve your chances of success. You can use the sample logs supplied in this book, other forms you have come across or developed, or a system such as the *SANY* software program.

Remember, your initial plan is not set in stone. You can change your goals if you need to. And when you reach your first weight loss goal, you can set a new one.

SELF-MONITORING

Self-monitoring is an important aspect of goal setting. What does it mean? It simply means keeping track of your own behavior.

People who are successful in losing weight and keeping it off find self-monitoring to be very helpful. This strategy involves keeping track of and recording:

- the foods and amounts you eat,
- the physical activities you perform, and
- regular measurements of your body weight and waist size.

Record the types and amounts of food you eat during your weight loss program using the diet log. Keep track of the amounts and types of physical activity you perform using the activity log. Sample logs and the Activity Table required to complete them are found at the end of Part II.

When you write all this down, it helps you pay attention to the changes you are trying to make. For example, you might be using a low-fat diet to lose weight and you might be keeping track of the fat grams you eat. Alternatively, you might be keeping track of the total calories you eat. Both are good ways to stay focused on what you're eating and to be sure you're eating foods that are consistent with your diet.

Keeping track helps you see where you are doing well and where you are not. Accurate records provide the information necessary to help you modify your eating behaviors to be more consistent with your diet plan.

STRESS MANAGEMENT

Many people find that they can manage their food intake most of the time—but that they eat more when they are stressed. We all live busy, stressful lives and starting a weight management plan may add to your stress at first. Get ready—if you develop coping strategies now, you'll be prepared for those stressful situations when they come. Here's how to develop coping strategies.

Choose alternate behaviors for dealing with stress.

If your tendency is to eat when stressed, think of another behavior—for example, taking a walk—that could replace eating in response to stress. If walking is not feasible, think about having healthier foods around to eat during those stressful periods. Some people find it useful to learn meditation or other relaxation techniques to help them avoid overeating in stressful situations.

PROBLEM SOLVING

You will face challenges as you follow your weight management plan. Start to see these challenges as problems that you can solve.

You can start your problem-solving process now, by identifying personal barriers that might prevent you from sticking with your plan. Where do you think you are most likely to have trouble and what can you do about it?

Solve problems before they appear.

If you think about potential obstacles ahead of time, you will have a solution when the problem arises. Perhaps you have a problem eating too much at buffets. One solution is just to avoid buffets—but you might not always be able to. If you have to go to a buffet, you might eat before you go, so that you are less hungry, or choose a salad or some other food choice that fits into your diet plan.

You will constantly be faced with weight management challenges. See these as problems and yourself as a problem solver. Some people find it useful to make a list of potential problems and the solutions for those problems as they execute their weight management plan. Try it!

STIMULUS CONTROL

Stimulus control is a very important behavior modification tool.

A stimulus is something that triggers a behavior. For example, seeing that the clock says it's noon may be a stimulus that tells you to eat, since you always have lunch at noon. Similarly, having a package of cookies in your cupboard might be a stimulus for you to eat the whole package.

Be aware of which stimuli lead you to engage in undesirable behaviors (that is, actions that put you in positive Energy Balance) and which stimuli lead to desirable behaviors (actions that contribute to your negative Energy Balance). Your challenge is to eliminate those stimuli that lead to undesirable behaviors and expose yourself to more stimuli that lead to desirable behaviors.

For example, it may be important for you to keep foods out of the house if they trigger you to overeat. These foods might be cake, cookies, ice cream, or any number of foods that are problem foods for you. If they are not in the house, there is less chance that they will cause you problems.

Stimulus control tips...

- Plan your meals in advance.
- Shop to match your meal plan.
- Don't skip meals.
- Use a small plate.
- Take small portions.
- Take at least 20 minutes to eat a meal.
- Wait at least 2 minutes before going back for seconds.
- Put your fork down between bites.
- Plan your physical activity in advance.

If eating at your desk at work is a problem, you might want to stop keeping food at your desk. Or you might want to keep healthful foods, such as fruits and vegetables, around the house or at your desk. Pick foods that you like and that can serve as reasonable replacements for your problem foods.

COGNITIVE RESTRUCTURING

Cognitive restructuring will help you replace your negative thought patterns with positive ones.

Your weight management plan is a long-term strategy. Let's face it; sometimes you will eat things that are not part of your diet plan.

Sometimes you won't meet your goals for your exercise plan. When that happens, it's easy to become discouraged and think negatively. Because you ate a box of doughnuts for breakfast, you might use that as an excuse to drop your weight loss efforts entirely.

Cognitive restructuring helps you turn this kind of negative thought or emotion into a positive one. These lapses are a natural part of making a lifestyle change. Accept what you have done, learn from it, and keep going. You may have learned what triggered that 12-doughnut binge, so now you're able to deal with that stimulus more effectively. Next time you'll use one of your behavior modification strategies.

Replace negative thoughts with positive ones!

You can accept what happened and modify the balance of your diet for that day by eating healthier foods. If you miss your planned physical activity one day, you can just make up for it by walking more over the next few days. Turn the negative into the positive and use every opportunity to learn about and understand your behavior.

SOCIAL SUPPORT

Weight management can be easier with a little assistance from our friends. Make the people who are around you most aware of your goals.

People are more likely to succeed with weight management when they have their spouse, family, or friends helping. Many people also find it useful to join a weight loss group to get additional social support. Sometimes it helps to be part of a group where other people are experiencing what you are going through.

WHEN YOU'VE FINISHED LOSING WEIGHT

How do you know when you are finished losing weight?

Perhaps you've reached your weight loss goal and that signals you to switch to your weight maintenance program. Some people, however, will not be able to reach their weight loss goal—they must decide whether to keep trying to lose weight or whether to concentrate on keeping off the weight they have already lost.

If you have reached your weight loss goal or feel that you have lost all of the weight you can (at least for now), then you might be ready to move from the weight loss phase to the keeping-the-weight-off phase. You can always decide later to go back to your weight loss program and lose more weight.

Even if you lose less weight than you had planned to lose, it is important for you to keep it off. The next chapter will help you make a plan for keeping your weight off.

CHAPTER 6. WEIGHT MAINTENANCE

James O. Hill, Ph.D.

Losing your weight is only half the battle. It is time to consider the other half—keeping the weight off.

This is where most people fail in weight management. It is natural to relax when you reach your weight loss goal and think that the hard part is over. But it is all too easy to fall back into old patterns of overeating and engaging in too little physical activity.

But stay aware! You know what will happen to your weight if you return to your old lifestyle. Many people who try to lose weight are successful at doing so. Very few are successful in maintaining that weight loss long-term. If you want to stay at your new reduced body weight, you need to continue with the Energy Balance concept and maintain the lifestyle changes that are consistent with your weight maintenance goals.

Think of weight maintenance not as the end of your program, but as the beginning of a new phase. You're moving from one phase, weight loss, to another phase—weight maintenance. You need different strategies for each phase.

We have shown you how to choose a strategy for losing weight and now we are going to help you choose a strategy for keeping the weight off. Now is not the time to let up. Celebrate your success, but realize there is still a lot of hard work ahead.

PERSONALIZATION IS THE KEY

We cannot tell you exactly what you need to do to keep your weight off. Just like with weight loss, you have to find the strategy that works for you. You are a unique individual, with a unique gene pattern and a unique lifestyle. You cannot do anything about your genes, but you can alter your lifestyle to keep the weight off. You have to personalize what will work for **you** in this phase, just as you did in the weight loss phase.

You may be asking why you cannot just keep doing what you were doing to lose weight. In fact, maybe you can. It really depends on how you lost the weight. The objective is to find a lifestyle that you can stick with forever. The problem with many diets is that, to lose weight, you make changes to your lifestyle that you cannot maintain indefinitely. You can lose weight on all of the popular diets, but can you eat like that permanently?

If you are thinking about staying with the dietary and physical activity patterns you used to lose weight, ask yourself, "Can this be a permanent way of life for me?". If the answer is yes, fantastic! You have your strategy. If the answer is no, then you need to consider how you are going to live the rest of your life.

SHIFTING FROM WEIGHT LOSS TO MAINTENANCE

It is very likely that your strategy for keeping weight off will be different than your strategy for losing weight. This is because there are some important differences between losing weight and keeping weight off. The table on the next page summarizes some of the most important differences.

During weight loss, your goal was to have negative Energy Balance, where your energy intake was less than your energy expenditure. This allowed your body to use stored energy (mostly body fat) to make up the difference between your energy expenditure and your energy intake. The greater your negative Energy Balance, the more weight you lost.

Differences between weight loss and weight maintenance

Weight loss	Keeping weight off
Negative Energy Balance	Zero Energy Balance
Temporary lifestyle changes	Permanent lifestyle changes
Your metabolism is decreasing	Your metabolism is not decreasing
Diet is most important	Physical activity is most important

You cannot stay in negative Energy Balance indefinitely. It is just not possible for people to be in negative Energy Balance for long periods of time. In fact, most people will be unable to maintain negative Energy Balance longer than about six months. You can expect to achieve most of your weight loss during the first six months of your diet.

DIFFERENT MIND SET

Keeping weight off may require a different set of skills and a different mind set. Your goal now is to achieve zero Energy Balance. It is no longer necessary for your energy expenditure to exceed your energy intake. The two just have to be equal. If you ensure that your energy intake and energy expenditure are equal, it is impossible for you to gain or lose weight.

While you might assume that achieving zero Energy Balance is easier than achieving negative Energy Balance, that is not necessarily true. Some people find the structure of a restricted diet easier to follow (in the short-term) than a less structured diet where they perceive they have more choices. The two phases of weight management are different and require different skills.

SUSTAINABILITY

A second difference is one that we have previously mentioned. Losing weight is a short-term task, while keeping weight off is a permanent undertaking. Many people can make substantial changes in their lifestyle temporarily—but it is much more difficult to make permanent changes in lifestyle. That's why we see that so many different strategies work for weight loss but few work for keeping the weight off.

METABOLIC CHANGES

Another difference between the two phases of weight maintenance is that, while your metabolism is working against you during weight loss, during weight maintenance it no longer works against you.

Remember your Energy Balance basics from Chapters 1 and 4. Resting metabolic rate, or RMR, changes with your body size. As you lose weight, your RMR goes down. Additionally, since the energy you expend in physical activity depends on body size, that also goes down as you lose weight.

During weight loss maintenance, you no longer have to worry about your metabolic rate going down as a result of your weight loss. In the weight maintenance portion of your program, your metabolism remains stable. That should make things easier.

INCREASED IMPORTANCE OF PHYSICAL ACTIVITY

A third difference between the phases of weight management is that, during the maintenance phase, physical activity takes on a larger role. During the weight loss (or dieting) phase, limiting energy intake was the most important factor in producing negative Energy Balance. During the comparatively short dieting phase, it was relatively easy to cut your energy intake by 500 or even 1,000 calories per day. Most people find it almost impossible to increase physical activity enough to achieve the same result. Basically, negative Energy Balance and weight loss is driven primarily by reducing energy intake.

The situation is very different when you want to keep weight off. Physical activity becomes as important as, if not more important than, diet. The people who succeed in keeping the weight off are the ones who make sure they get regular physical activity.

Physical activity is the key to weight maintenance.

In summary, you cannot expect the strategy that produced weight loss to keep the weight off. Perhaps so many people regain lost weight precisely because they expect to keep the pounds off by staying on their weight loss program.

Most people do succeed in making the temporary changes needed to lose weight—but they go into weight loss with the expectation that they will maintain these changes permanently. When they find that they are not able to sustain the changes, they feel they have failed. Usually they respond by returning to the patterns of eating and physical activity that caused them to gain weight in the first place.

What's missing for so many people is a way to transition from weight loss and negative Energy Balance to weight maintenance and zero Energy Balance. Here's how it works.

THE IMPORTANCE OF TRANSITION

To consider how to handle this transition, let's go back to our Energy Balance equation. Assume you were in zero Energy Balance before you started to lose weight. You may have been overweight or obese, but you were not gaining weight. Now that you have lost weight through negative Energy Balance, you may think you can you return to your previous lifestyle. After all, it kept you in zero Energy Balance.

You can't! Here's why.

Because you've lost weight, you now have a smaller body mass. Your new, smaller body (congratulations!) requires less energy to operate—you have a decreased total metabolism. You have fewer cells to which energy must be supplied, a smaller circulatory system through which blood must be pumped, and so on. If you go back to the same physical activity and eating patterns you had before you began your weight loss program, you will regain the weight you lost.

Fortunately, that doesn't have to happen. It is possible to estimate your new energy requirements and to determine what to do differently to achieve and maintain zero Energy Balance in your smaller body.

First, let's make sure you understand the mathematics of what happened during the weight loss process and the implications of that for a successful maintenance program. The following example illustrates the cycle of weight loss and weight gain that typifies the experience of the estimated 80% to 95% of dieters who fail to maintain their weight loss.

THE WEIGHT LOSS, WEIGHT GAIN CYCLE

Assume that you weigh 200 pounds and that your resting metabolic rate (RMR) is 1,600 calories a day. On top of that, you burn an extra 650 calories per day in physical activity. By adding these values together, we determine that during a 24-hour period your body burns 2,250 calories.

Before and after caloric requirements

Description	Before diet	After weight loss
Weight in pounds	200	170
Calories burned by RMR	1,600	1,500
Calories burned in activity	650	600
Total calories expended	2,250	2,100

You begin your diet, it is successful, and you lose 30 pounds. First of all, this is a 15% weight loss—which is very good and should improve your health and quality of life significantly. Now you weigh 170 pounds and your resting metabolic rate has declined from 1,600 calories a day to 1,500 calories a day because you have a smaller body. You do the same amount of activity as you used to, but because your body size is smaller, it burns, for example, only 600 calories instead of the 650 calories each day it burned when you weighed 200 pounds. Your daily energy expenditure has been reduced to 2,100 calories.

Notice that your RMR does not go down in proportion to your loss of body weight. In fact, it goes down less. This is because when you lose weight, most of your weight loss is from fat and not from lean body tissue or muscle. It is the lean tissue that contributes to RMR and, since only a small amount of your weight loss was from lean tissue, your RMR has not decreased in proportion to your body weight.

If you were in zero Energy Balance before weight loss, you were eating 2,250 calories and burning 2,250 calories. After weight loss, you're burning 2,100 calories. If you go back to eating 2,250 calories, you will create a positive Energy Balance of 150 calories per day because your food intake will be greater than your energy expenditure.

As you recall from Chapter 1, it takes only about 10 calories a day of positive Energy Balance for you to gain a pound every year. At that rate, you will regain weight until you reach your previous weight of 200 pounds and your metabolism returns to 2,250 calories. You will again be in zero Energy Balance—but you will have gained back all your weight. This is the cycle of weight loss and weight gain that most dieters go through over and over.

BREAKING THE CYCLE

You can use your new understanding of Energy Balance to break the cycle of weight loss and weight gain. If your energy expenditure after weight loss is 2,100 calories per day, then you can choose from a number of options for achieving zero Energy Balance:

- You could eat 2,100 calories per day (150 calories per day less than before weight loss) and keep your physical activity the same as it was before you lost weight. This will put you in zero Energy Balance at 2,100 calories per day.
- Or you can eat the same as you did before you lost weight (2,250 calories per day) and increase your physical activity to burn an additional 150 calories per day. This will put you in zero Energy Balance at 2,250 calories per day.
- Finally, you can eat a little less and increase physical activity a little more. Depending on how much change you make in each, this will put you in zero Energy Balance somewhere between 2,100 and 2,250 calories per day.

Adjust for your new metabolism.

Once you define the problem in Energy Balance terms, you see that you have choices in how to solve the problem. Some people will find it easier to eat a little less and some will find it easier to increase physical activity a little more. Let's assume you decide to split the difference and eat 75 calories less each day and increase your physical activity by 75 calories each day. Both of these goals are readily achievable.

UNDERSTANDING YOUR "ENERGY GAP"

The difference between your energy expenditure before you lost weight and your energy expenditure after you lost weight is your **energy gap**. Your energy gap represents the number of calories you need to eliminate through some combination of eating less and exercising more.

Energy Gap = Output Before Weight Loss - Output After Weight Loss

---

Because everyone's physiology is unique, the extent of each person's drop in energy expenditure with weight loss is also unique. There is no easy way to calculate your energy gap accurately without the aid of a sophisticated program like the *SANY* software. However, estimating your energy gap as 5% to 10% of your energy expenditure before you lost weight should be close enough to allow you to plan a useful strategy.

The problem is very straightforward when you look at it in terms of Energy Balance. The table below illustrates the energy gap calculation.

Closing your energy gap

Description	Before diet	After weight loss
Weight in pounds	200	170
Calories burned by RMR	1,600	1,500
Calories burned in activities	650	600
Total calories	2,250	2,100
Energy gap	n/a	2,250 -2,100 =150
Decrease food intake	n/a	-75
Increase exercise	n/a	-75
Remaining energy gap	n/a	0

This knowledge will help you estimate the amount of calories required by your new RMR and your activities. The calculation will help you decide how much to change your diet and how much to change your physical activity.

ESTABLISHING YOUR ZERO ENERGY BALANCE

You may want to give some thought to the level at which you want to establish your zero Energy Balance.

Yes, you do have a decision to make, and it is an important one. Should you close your energy gap by reducing your calorie intake, or should you close the gap by expending more calories through increased physical activity?

If you close your energy gap by eating less, you'll reach zero Energy Balance at a smaller number of calories. If you close it by increasing your physical activity, you'll reach zero Energy Balance at a higher number of calories.

There may be an advantage to choosing the higher Energy Balance level—that is, the option of increasing your physical activity. The reason is simple and a direct result of our genetic heritage. Humans are by nature not good at restricting food intake over long periods of time. Therefore, the more calories you can eat without gaining weight, the less you'll experience feelings of restriction and deprivation. That's one of the reasons we recommend you emphasize physical activity during weight maintenance. The *SANY* software can be of great assistance in this because it provides the accurate feedback required to efficiently manage your Energy Balance.

For every 100 calories you burn through exercise, for every mile you walk, you can eat 100 calories more. While physical activity can't provide the high caloric burn rate needed to lose weight quickly, it is ideal for burning smaller amounts of energy on a regular basis.

And there is an additional benefit to regular physical activity. As you increase your physical activity, you increase your muscle mass. Since muscle burns more energy per pound of body weight than fat does, your exercise will be increasingly effective.

Consider using physical activity to make up as much of your energy gap as possible.

SUCCESSFUL WEIGHT LOSS MAINTAINERS

We can get some great tips for how to succeed in keeping weight off from the people in the National Weight Control Registry. Remember these folks? They've been able to maintain an average weight loss of 67 pounds for an average time period of 6 years.

You read earlier that the only similarity in how these individuals lost weight was that they used both diet and physical activity. There was very little similarity in the kind of diet they chose or in the manner in which they increased their physical activity.

That diversity disappears when we look at how these people successfully keep their weight off.

Four common behaviors in successful weight maintenance:

- Regular self-monitoring.
- Eating a low-fat, high-carbohydrate diet.
- Eating breakfast every day.
- Engaging in high levels of physical activity.

SELF-MONITORING

We talked about self-monitoring earlier. Self-monitoring means keeping track of your weight, food intake, and physical activity.

People in the National Weight Control Registry weigh themselves frequently. Most weigh themselves at least once a week, and many do so several times a week. Similarly, they regularly record their food intake and physical activity. They write all this information down in their diet and physical activity logs.

Keep track of your weight, your energy intake, and your energy output.

Why is self-monitoring so important? Weighing yourself frequently allows you to see if you are gaining weight before you gain too much. It serves as an early warning system. Rather than waiting until you've regained ten pounds, you can start doing something about it when you see that you have gained one or two pounds. You can make small adjustments to your diet or activity levels before you have gained so much weight that you become discouraged and give up on your weight maintenance program.

It is also important that you know what you are going to do if you do gain a pound or two. Self-monitoring a behavior, such as food intake or physical activity, makes you pay attention to that particular behavior. It is important that you take the time to keep track of your behaviors.

Self-monitoring keeps you in touch with your body's energy needs so that you can make small adjustments as they become necessary.

DIET

The second thing that these successful weight loss maintainers have in common is a diet low in fat and high in carbohydrate. They report that about 24% of their total calories come from fat, 20% from protein, and 56% from carbohydrate.

A low-fat diet may be more important in preventing weight gain than it is in producing weight loss. Scientific research shows that when people are allowed to eat all the food they want, they consume more total calories on a diet high in fat than on a diet low in fat. Additional research suggests that low-fat diets may be particularly effective—at least for some people—in preventing weight gain[15] [16].

---

[15] Lissner, L., & Heitmann, B. L. Dietary fat and obesity: evidence from epidemiology. *European Journal of Clinical Nutrition*. 1995; 49:79-90.

BREAKFAST

The third behavior that these individuals have in common is that they are breakfast eaters. In fact, most of the people in the National Weight Control Registry report that they eat breakfast every day. Only 4% report that they never eat breakfast.

Eat breakfast!

It appears that skipping breakfast to manage weight is **not** a good idea. Eating breakfast may contribute to successful weight maintenance by helping manage hunger. Skipping breakfast may increase the likelihood of overeating at later meals.

PHYSICAL ACTIVITY

The fourth behavior common to people in the National Weight Control Registry is that they are very physically active. They report that they expend about 2,700 calories each week in physical activity. This is equivalent to 60 to 90 minutes a day of a moderate physical activity such as walking.

In fact, engaging in regular physical activity seems to be the best predictor of long-term success in keeping weight off. Only 9% of the 3,200 people in the National Weight Control Registry report that they maintain their weight loss without engaging in regular physical activity.

Physical activity
is the best predictor of success in keeping weight off.

---

[16] Hill, J. O., Melanson, E. L., Wyatt, H. T. Dietary fat intake and regulation of Energy Balance: implications for obesity. *Journal of Nutrition 2000*; 130:284S-288S

Why is physical activity so important for maintaining a weight loss? There are several reasons. Here are just four…

- Physical activity increases energy expenditure and helps close the energy gap created by weight loss.
- Most of the fuel needed for the increased physical activity comes from fat, exactly what you want to burn.
- Muscles of physically active people use more fat for fuel than muscles of physically inactive people.
- Physical activity stimulates changes in brain neurotransmitters and chemicals that may actually help you regulate your appetite.

Engaging in regular physical activity is associated with other positive health results. People who lead physically active lives report better overall health than physically inactive people. Increasing physical activity also has positive psychological effects. For example, it increases self-esteem and reduces depression.

Potential benefits of physical activity:

- Burns mostly fat.
- Produces changes in brain neurotransmitters that may help you regulate your appetite.
- Improves overall health.
- Reduces depression.
- Boosts self-esteem.

The human body was designed to work most efficiently at high levels of physical activity and when high energy intake matches high energy expenditure. Another way to look at this is that we were designed so that physical activity would be the driver and food intake the follower.

If you are not physically active, your energy expenditure is low. To maintain a zero Energy Balance, you have to restrict your energy intake to match your low level of energy expenditure. This is very difficult to do, primarily because, physiologically, we are designed not to restrict calories.

The best way to achieve zero Energy Balance—and keep weight off—is to incorporate regular physical activity into our lifestyles.

HOW MUCH IS ENOUGH?

We've established that physical activity is very important for keeping weight off—but how much physical activity do you really need?

The answer is—it depends. It depends on all of the factors that we've talked about: your genetic make-up, your lifestyle, and how much you are able to restrict your food intake. The one thing that's certain is this—maintaining your weight loss will require you to be more physically active than you were before you lost weight.

The Surgeon General of the United States recommends that people get 30 minutes a day of moderate intensity physical activity, such as walking at a reasonable pace or riding a bike at a reasonable pace. Thirty minutes is a nice, comfortable number that many of us think of as an acceptable amount of time for daily exercise.

Unfortunately, the evidence supporting the need for 30 minutes a day of physical activity is specifically related to the heart. That's good news for the health of our heart. Walking for 30 minutes a day is easy for most of us.

You may need 60 minutes a day of moderate exercise.

The bad news is that experts now believe that it takes more than 30 minutes a day of physical activity to optimize weight control. Experts now recommend 60 minutes a day of moderate intensity physical activity to keep from gaining weight. If you are maintaining a large weight loss, it may even take more than 60 minutes a day of physical activity for you to keep the weight off.

THE EFFECTS OF PHYSICAL ACTIVITY ARE CUMULATIVE

The good news is that you do not have to get all of your physical activity in a single bout. You can do 5-minute bouts, 10-minute bouts, or any combination that adds up to your goal. So, get in the habit of adding small bouts of exercise consistently throughout your day.

What is the best way to increase your physical activity? The best physical activity for you is anything you will do consistently. It's important that you increase activity rather than worrying about doing the perfect exercise.

The best exercise for you is the one you will do.

Walking is the most popular form of physical activity for people in the National Weight Control Registry. We have consistently found that walking is a great way to increase your energy expenditure and to help prevent weight gain. Consider adding several short walks to your daily routine.

INTENSITY DECREASES TIME COMMITMENT

An hour a day? How—you might be asking—can I possibly fit an hour of walking into my busy day?

Here's more good news. The higher the intensity of your physical activity, the more energy you expend during every minute that you're exercising.

Again, you have choices. If you want to get the greatest energy expenditure in the shortest period of time, then exercise at a higher intensity. If you decide to stay with the lower intensity, plan to exercise for a longer period of time each day.

You can choose from many physical activities: walking, running, swimming, cycling, rowing, weight lifting. It really doesn't matter. All of these activities burn calories and all really help in keeping the weight off.

Once you've set your goal for the amount of energy you're going to expend through physical activity, your next step is choosing your type (or types) of physical activity. Then calculate the duration for that activity that allows you to achieve your goal. The Activity List/Metabolic Rate Table in Part II will help you do this by showing you how to determine the energy you will expend through each type of physical activity based on your RMR. A sample log is provided in Part II to help you organize your data and make the necessary calculations.

However, the *SANY* software is probably the easiest, most efficient way for you to manage your Energy Balance. Not only will the software automatically calculate your energy expenditures based on your actual RMR, but it will provide you with your net Energy Balance at any point in time. Depending on your progress relative to your objectives, the *SANY* software will provide helpful suggestions based on your personal data that will make your program easier and more effective.

One other observation made by researchers studying the National Weight Control Registry is that the risk of regaining weight decreases after people have kept the weight off for two or three years[17]. At first, you'll have to work hard to keep your weight off, but this observation suggests that things will get easier. If you can keep your weight off for two or three years, there is a very good chance that you will continue to keep it off.

Remember, keeping weight off is just as important as losing the weight in the first place. The key to keeping that weight off is closing the energy gap. You can close the gap through any combination of increased physical activity and decreased food intake.

THE IMPORTANCE OF INDIVIDUAL DIFFERENCES

Throughout this book we have emphasized that there are big differences in how individuals achieve and respond to changes in their Energy Balance. If we put 100 people in the same degree of negative

---

[17] Wing

Energy Balance—let's say -500 calories per day—every one of these people will lose weight. There will, however, be large differences in how much weight each person loses. The reason for the variation is not totally clear, but it's thought to be related to differences in physiology caused by differences in genetic make-up.

We can illustrate this with a famous research study. Researchers at the Mayo Clinic fed 16 volunteers 1,000 extra calories each day for 8 weeks. Researchers lived in the research facility so that they could monitor exactly how much food each person ate[18]. Participants' physical activity was limited.

Despite the fact that each subject ate exactly 1,000 extra calories each day, the amount of weight gained differed widely among the subjects. In fact, the amount of fat gained over the 8 weeks differed 10-fold. These widely ranging results are caused by the unique way an individual's metabolism adapts to positive Energy Balance.

Similar differences would likely occur during negative Energy Balance. It is almost certain that, if we gave 100 people a negative Energy Balance of 1,000 calories, the number of pounds lost by the participants in our study would differ widely.

There are many possible explanations for this. For instance, RMR declines more rapidly in some people than it does in others. And some people spontaneously move around more than others do when they are in negative Energy Balance.

The bottom line is that, if you are in negative Energy Balance, you will lose weight. But it's impossible to predict, based on averages, exactly how much weight you will lose and whether your weight loss will be greater on one plan than another. We also cannot predict whether you will be more successful if you emphasize diet more or if you emphasize physical activity more.

We know that all of these variables exist. The need to predict how individuals will respond to changes in their Energy Balance stimulated the development of the *SANY* software described in Part II. Part II also

---

[18] Levine, J. A., & Jensen, M. D. Role of nonexercise activity thermogenesis in resistance to fat gain in humans. *Science magazine.* 283:212-214.

gives you the tools to personalize your weight management program, to make it more effective and help you keep your weight off—permanently.

Part II

PERSONALIZING WEIGHT LOSS

SANY SUPPORT SYSTEMS

The book *Shaping a New You* is paired with an important new software program designed to assist you with your weight loss and maintenance programs. We have already discussed the importance of record-keeping and self-monitoring. This software helps you through those important tasks and provides you with crucial information about your body's unique physiology.

SANY– A NEW DIMENSION

This new software, which is also called Shaping a New You, or *SANY*, is your virtual personal weight loss and weight maintenance coach. The *SANY* software guides you through your program and helps you keep your records accurately, easily, and efficiently. *SANY* performs all the calculations that you would normally be unable to make accurately: it calculates your energy gap, establishes your zero Energy Balance, and tells you your personal RMR. It makes recommendations for your diet and activity plans that are customized to your personal data, provides you with accurate and essential feedback, and even lets you try "virtual" diets to see how they might interact with your unique physiology and preferences.

SEE THE RESULTS BEFORE COMMITTING

The *SANY* software will also show you, in three-dimensional graphics, how your body is likely to look after you have made proposed changes to your diet or activity levels. You can get a sneak preview of how your body will have changed after six months (or however many months you choose) on your program. *SANY* takes the guesswork out of your weight loss and weight maintenance.

As Dr. Peters pointed out in Chapter 1, our failure to win the war on weight does not mean we have a national case of poor willpower. Weight loss is a complex process. To lose weight, we need to manage our energy intake and energy expenditure in a way that accommodates our unique physiologies.

We know the principles of weight loss, but translating them into practice is not easy. For example, how do you effectively monitor your entire day's activities, record your food intake, and assess the net effect when a difference of 10 to 50 calories a day could make the difference between weight gain and weight loss? How can you achieve that degree of accuracy in your record-keeping?

Accurate feedback is essential.

Without the understanding of Energy Balance that you now have, it is extremely difficult for most people to develop an achievable, realistic weight goal. You've already successfully set your weight goal. *SANY* will show you how to translate your weight goal into Energy Balance terms. Empowered with this information, you will know what to do to achieve your desired weight, how long it will take you to lose the weight, and how to keep the weight off once you've lost it.

Now that you've defined your task, consider what tools you have to do the job. You need tools that will help you understand your own Energy Balance and that will help you analyze your current eating and physical activity behaviors, so that you can adjust these behaviors in a way that will let you meet your weight goal.

Picking the right tools is important.

You can find tools to help you with your Energy Balance goals at bookstores and on the Internet. It may be helpful, for example, to buy a book with the nutrient and energy content of a variety of foods. This will help you to estimate the total energy in the foods you eat each day. Similarly, you can buy a book that lists the energy burned during various types of physical activity so that you can estimate your total energy expenditure based on your records of the physical activities you perform each day. But this is painstaking, time-consuming work—the *SANY* software can eliminate this work by doing it for you.

There are also dieting services online, which allow you to subscribe, enter the foods you eat and the activities you perform, and calculate energy intake and energy expenditure based on averages. But these averages may not reflect your personal physiology and may lead you to make decisions that just don't work for you.

SANY can make the whole process a great deal simpler. The artificial intelligence software developed for use with this book is the most advanced weight management tool there is. This software not only allows you to calculate your energy intake and energy expenditure from your recorded food intake and physical activity, but it also spots complex patterns in your past behaviors that can be useful in helping you modify your future behaviors. In other words, *SANY* makes its calculations **based on your personal physiology and behaviors and it suggests changes based on your physiology and behaviors**.

APPLYING TECHNOLOGY

Clearly, managing our Energy Balance is a complex undertaking. We have developed technologies that help us understand and manage complexity, and others that provide effective, simple ways to solve problems, in many aspects of our lives. The pedometer, heart rate monitor, and weight scales are all examples of technologies that can be used in weight and fitness applications. These are examples of relatively simple applications where just one or two inputs are measured.

Each of these technologies provides important information that can help us manage our weight—but they fail us, because they don't help us integrate all that valuable information and they don't place all that information in the context of our unique circumstances.

As we have seen, no two people respond in the same way to any weight management plan. Weight loss and weight maintenance are not "by the averages" processes. Dealing with the numerous elements that contribute to weight gain demands a system that can understand cause-and-effect relationships and that can learn from an individual's results over time. *SANY* is such a system.

Until now, no system has been able to integrate all those complexities into a coherent, individualized weight loss and management program that learns about your individual physiology and responds based on that individuality.

But we now have the system we need--*SANY.*

MANAGING THE UNMANAGEABLE

As our weight management experts have explained in this book, weight loss is a function of Energy Balance. Energy balance is an easy concept to understand—and it is critically important to weight management. This Energy Balance concept goes beyond the elements of any single diet plan.

You now understand some of the subtle quirks of weight loss, especially the decrease in metabolism that accompanies the loss of weight. You know that successful weight loss can be the result of very modest changes in energy intake. You know that your environment will provide you with unlimited temptation, which you can fight using a combination of understanding and willpower.

It is one thing to understand the concepts involved in successful weight loss and maintenance. It is quite another to make accurate calculations, tie the individual category results together (e.g., waist measurements over time, weight, calories consumed, and so on) into a unified understanding of what your body is doing. *SANY* ties all these bits of data together to produce an integrated, personalized program for you.

HARNESSING A STATE-OF-THE-ART TECHNOLOGY

SANY will assist you in achieving your desired weight by personalizing your diet and fitness plans. It will even take into consideration your current health status. As you use *SANY*, it continually learns about your dietary and activity patterns, your likes and dislikes, and your physiology. It then customizes its projections to you, as a unique individual, and builds your personalized fitness and diet programs accordingly. If your health status or nutritional and physical needs change, *SANY* adjusts your programs and projections to match your new status and needs.

Personalization makes a huge difference!

The advantage that the *SANY* software provides over traditional approaches to weight management is that *SANY's* projections are both truly personalized and based on scientific principles. You are not given estimates based on averages of large numbers of faceless people in all sorts of circumstances. Instead, you are given precise information that is applicable to you and only to you and you can use it without constantly having to enter your personal information on some Internet site.

SANY identifies patterns in your food intake and physical activity over time and generates a diet program based on those patterns. These patterns are the result of very complicated interactions at the molecular level inside the human body, and they are different in every person.

It's one of the reasons traditional approaches to weight loss—which are based on averages—fail. The reaction to any given weight loss program will be different from person to person. Successful weight loss and management benefits from accurate, user-specific feedback—which, until now, has not been available.

SANY does not need to know the exact mechanisms that govern the physiology of the human body. The software matches the inputs you provide over time with the outputs generated to identify your individual patterns and helps you make the right decisions about how to accomplish your desired weight loss goal.

For example, two people with same anthropometric values (weight, height, age, gender, waist circumference, and so on) who are on the same diet will likely lose weight at different rates because of metabolic differences. *SANY* will take into account their unique physiologies (for example, how their body weights change over time) and provide both with equally efficient diet programs.

In order for *SANY* to work, you have to tell it about what you eat and the physical activity you engage in. This 'trains' the system so that it can provide you with feedback that has been tailored to fit you.

The advanced *SANY* system can do this even when the information you provide is incomplete. This is an important feature—because even the most organized person occasionally forgets, or is too busy, to input their diet and physical activity data. Once the software begins learning about your preferences, it will help you fill in the required daily logs based on your previously entered eating and activity patterns.

HOW DO YOU FEEL?

The software also asks you every week or so to tell it how you are feeling about your weight management plan. You might say, for example, that you are very hungry, somewhat hungry, or not hungry at all. By taking into account your hunger level, *SANY* modifies your weight management program to help maximize your sense of well-being and comfort.

SANY also generates a personalized fitness program for you based on what you've told it. At first, it will present you with a set of exercise choices. Once it knows your preferences, it will adapt the exercise list to include only those you have indicated you might do. You will, of course, be able to update the list if your exercise preferences change.

THE "WHAT IF?" MODULE

The *Shaping a New You* software allows you to "see the future you," in 3-D at any time, as you progress in your personal program. *SANY*'s powerful artificial intelligence system allows you to perform "What if?" scenarios and see the probable results.

For instance, what would be the effect on your weight if you stopped drinking three sodas a day? How would that translate into weight loss over a six-month period based on your metabolism and activity levels?

What if you could predict your results?
YOU CAN.

SANY will show you the projected effect, in 3-D, based on your current measurements, RMR, activities, gender, and weight. You can perform any number of "What if?" scenarios to find the right combination of diet and activity for you.

Think of being able to try out a new option without having to commit time and effort—until you know what effect it will have on your weight! This "What if?" imagery is a powerful and unique tool, and it's available only in the *SANY* software.

FOR THE BEGINNER AND THE ADVANCED USER

The *SANY* application does not assume that you are an expert on diet and fitness. The application has a user-friendly interface, with an interactive "Wizard" that guides you through it.

The *SANY* system has numerous advantages over other weight management products on the market.

For instance, it:

- Provides the ultimate in individualization, including the "What if?" module.

- Provides the personalized feedback necessary for successful weight loss; this is feedback that couldn't be produced manually without extensive calculations.
- Retains your personal information on your computer, not on the Internet.
- Simplifies the process of record-keeping and input.
- Creates both visual and graphic feedback that is easy to understand.
- Gets smarter as you give it more information.
- Allows you to pick a diet that fits your lifestyle and habits.
- Provides a nutritional evaluation of your eating habits and makes recommendations based on your likes and dislikes.
- Uses the very effective concept of Energy Balance.
- Eliminates the need for monthly membership fees.

SANY is a lifetime system that can be effectively used to monitor your dietary and physical activity levels even after you have completed your weight loss program.

Remember, the tendency for most of us is to gain small amounts of weight over time. *SANY* can help you avoid that trap.

PERSONALIZING YOUR WEIGHT LOSS PROGRAM

James O. Hill

The first six chapters in this book gave you important information about the concept of Energy Balance that will allow you to take charge of your body weight. Now it is time for you to put this information to work. Part II of this book is designed to help you apply your new knowledge about Energy Balance.

We believe we have given you the true story about weight loss. It is not about a "magic bullet." It is all about numbers and commitment. If you understand this, you have a great advantage over most dieters.

Let's review what you know:

- You know that, to lose weight, you need to produce a negative Energy Balance, and that you can do so through a combination of eating less and moving more.
- You know that the transition from weight loss to weight maintenance is an important one and that, for successful weight maintenance, you have to achieve and maintain zero Energy Balance.
- You know that few people succeed in achieving and maintaining weight loss through diet alone. Successful weight maintenance also requires an increase in physical activity.
- Finally, you know that different strategies for weight loss and weight maintenance work for different people.

Consider yourself ahead of the game. You have the truth about weight management.

Now we come to the tough part—determining your own unique Energy Balance. It's a difficult calculation, because you have to keep track of everything you eat and then convert all the foods you've eaten to calories consumed. You also have to keep track of all the physical activities you perform and convert those to calories burned. It is hard to do all this with a high degree of accuracy unless you use a system such as the *SANY* software. Look at the *SANY* screen below.

Portion of *SANY's* main screen

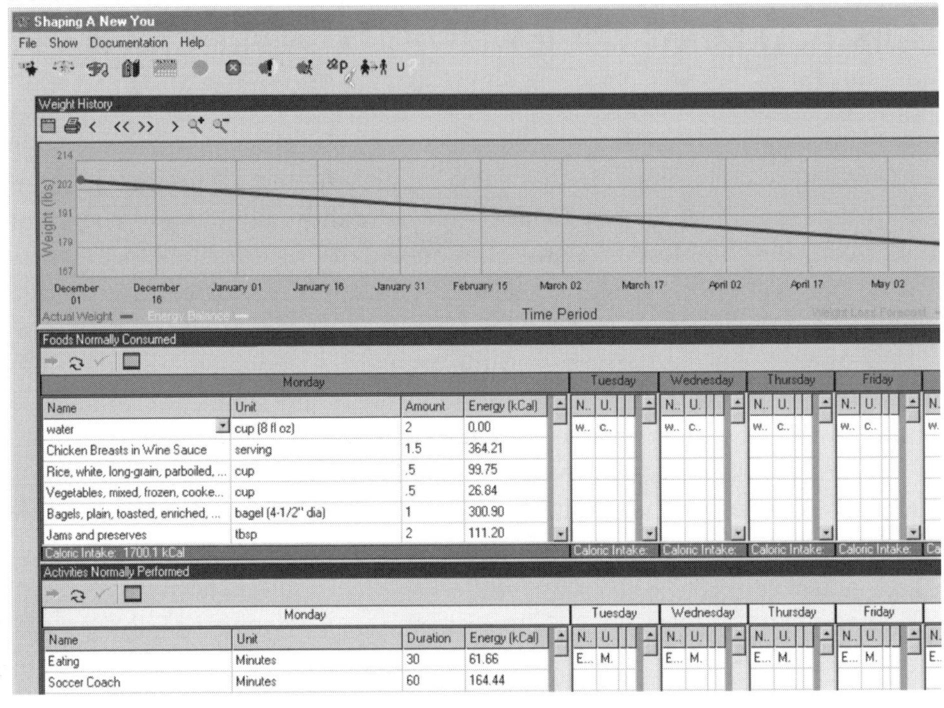

This screen shows you a sample weight loss program. The diagonal line represents the user's weight loss goal over time. The *SANY* software makes the necessary calculations to determine if the person is on track, above, or below the desired results—based on the weight goal, diet information, and activity information entered by the user—and gives the user customized feedback.

Record-keeping is made easier by the *SANY* artificial intelligence system, which will fill in many of the "blanks" once enough background information has been entered. The software allows the user to make changes to the data at any time.

The *SANY* software's objective is to keep the user on the track to achieving their weight loss goal. If the dieter loses weight too quickly, the system may suggest increasing caloric intake so that the diet is not overly restrictive. If activity levels are low, the system will make suggestions based on the individual's preferences.

SANY constantly monitors nutritional adequacy and provides an easy to understand graphic representation of the user's diet based on USDA recommendations. For example, if sodium levels are too high, *SANY* will provide suggestions on which foods to avoid to reduce sodium intake. *SANY*'s feedback is color-coded to make it even easier to understand.

***SANY* nutritional feedback window**

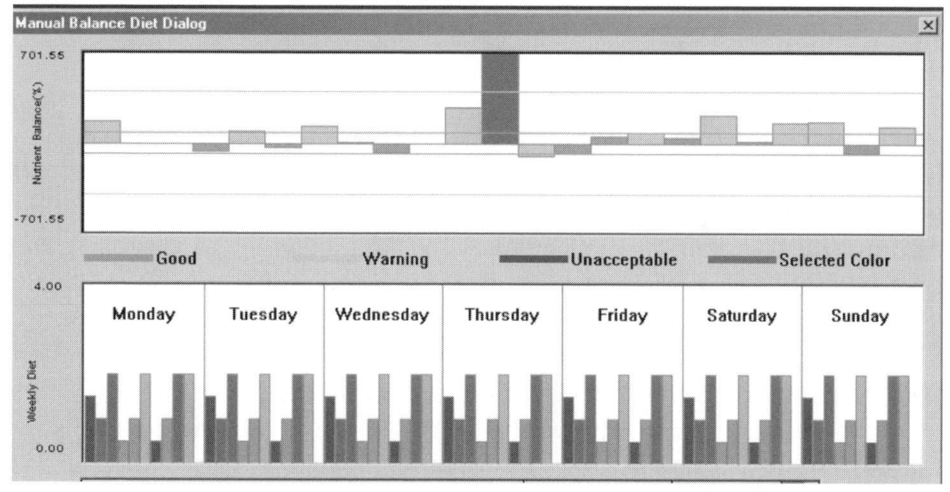

SANY provides information that will simplify the user's weight management program and render it as effective as possible. The 3-D image generator is an excellent example. Using the 3-D image generator with the "What if?" module, a person can see the probable results of any proposed change to their weight management program. Simply enter the dietary or physical activity change into the system, including the desired time frame, and the results will be displayed based on the user's unique metabolism and personal data.

3-D simulation can show either male or female images

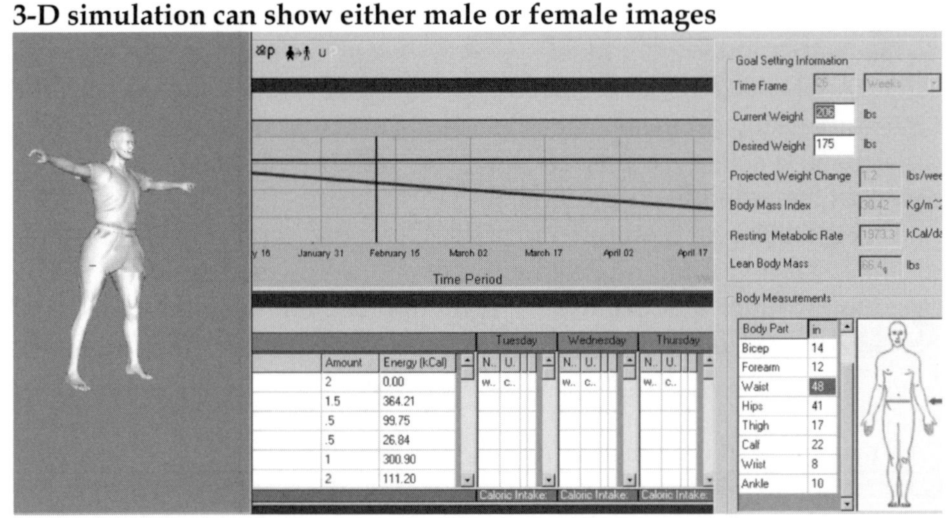

Remember our discussion about your "energy gap" in Chapter 6? The energy gap is the difference between your energy expenditure (or output) before you lost weight and your energy expenditure after you lost weight. We suggested a way you could estimate that gap, but we didn't recommend that you try to calculate your actual energy gap, because the calculation is extremely complex. But the *SANY* software can calculate your energy gap precisely and incorporate it into your daily Energy Balance results. This gives you an exact picture of how you're doing on your weight management program on any given day.

These examples illustrate just a few of the many advantages of using the *SANY* software. The *SANY* software makes complex calculations effortlessly and provides you with easy-to-understand feedback presented in graphic and in text.

If you decide not to use the *SANY* Artificial Intelligence software, you can still use the Energy Balance concepts you've learned in this book and perform your own calculations. We'll describe the basic calculations in the following paragraphs.

There are several steps to the calculation:

- First, determine how many calories you are currently consuming. There are books that give calorie values for just about any food you might eat. You can use the food logs we provide to keep track of what you eat.
- Use a calorie book to convert your logged foods into energy intake.
- You can also determine how many grams of protein, carbohydrate, and fat you are currently eating.
- Then you can periodically keep food records while you are losing weight to check if you are meeting your energy intake goal.

While it does take effort to track your food and convert it into calories, it is clear that people who do so are more likely to be successful weight managers than people who don't.

You can estimate your total energy expenditure by doing similar calculations of your physical expenditures. We have told you how to get your RMR at the Shaping a New You website. Now you can estimate how much energy you expend in physical activity each day.

There are different ways to do this calculation. The easiest (but perhaps the least accurate) is to multiply your RMR by a factor that represents your average activity level over the course of a day.

MEASURING PHYSICAL ACTIVITY

If you get almost no physical activity, your daily energy expenditure is your RMR x 1.2. In this case, 1.2 is your **general activity factor**. If you get a little activity but are not regularly active, use 1.4 as your activity factor. If you get regular exercise on most days, use 1.6. Finally, if you are active almost all day and have periods of high activity, use 1.8.

General estimation of energy expenditure

Level of physical activity	Multiplier factor
Very sedentary	1.2
Generally low level of activity	1.4
Regular activity and some strenuous exercise	1.6
Active most of the day, periods of high energy output	1.8

A second, more accurate alternative for estimating how many calories you burn in a day is to use the physical activity log to keep track of your daily activities, including sleep periods. Look at the table below.

Sample activity log

Activity	Metabolic rate factor	RMR	Time in minutes	Divide by 1440	Calories burned
Sleeping	0.9	1457	480	1440	437
Office work	2.5	1457	300	1440	759
Sitting (meeting)	1	1457	180	1440	182
Standing	1.2	1457	120	1440	146
Baking	2	1457	60	1440	121
Driving	3	1457	90	1440	273
Aerobics	6	1457	65	1440	395
Watching movie	1	1457	120	1440	121
Other	1	1457	25	1440	25
Total minutes should be 1,440			1,440		
Total calories burned					2,460
Energy input from food and drink (from food log)					3,266
If calories burned exceed energy input, you have achieved a negative Energy Balance of:					
If energy input exceeds calories burned, you have achieved a positive energy Balance of:					**806**

If you know how much time you spend doing each of your daily activities, multiply this time by the energy value shown for that activity. The sum is an estimate of the energy you expended in physical activity during the day. Because the calculation demonstrated in the above table takes your RMR into account, it's much more accurate than, for example, the exercise chart illustrated in Chapter 4 or the general estimate procedure described above.

It takes more time to achieve this level of accuracy, but your weight management program will be far more effective if you take the time to use this type of log.

Here is how you use the log:

- Find the activity on the Activity List/Metabolic Rate Table following this explanation. If your activity isn't listed, pick an activity that is similar.
- Enter the **Metabolic Rate Factor** found in the column to the right of each activity. For example, sleeping has a metabolic rate factor of 0.9.
- The **RMR** will be the same for all daily entries. Check your RMR about once every week or two and adjust your calculations accordingly. The Shaping a New You website (www.shapinganewyou.com) has an easy-to-use calculator that will give you an estimated RMR. In the example above, the RMR is 1457.
- Enter the time (in minutes) that you spent in each activity. Be sure to include sleeping and other sedentary activities. The calculation is based on 24 hours of activities; it will not be accurate unless you account for all 24 hours of the day. The example illustrates 480 minutes of sleep (8 hours).
- Calculate the calories burned (energy used) for each entry using the formula explained below.

EXAMPLES

The formula used to calculate energy expenditure for your daily activities is:

Calories Burned=[**Metabolic Rate Factor** of an activity] x [**Your Estimated RMR**] x [**Minutes Doing Activity**]. Divide the result by **1440** (minutes in a day).

Examples:

Calories burned sleeping=(.9 X 1457 X 480)/1440=437 calories burned

Calories burned during office work=(2.5 X 1457 X 300)/1440=759

Activity list / Metabolic rate table

Activity list	Metabolic rate factor
Aerobics (moderate)	6
Archery	3.5
Assembly work	3
Attending class	1.8
Attending meetings	1.5
Backpacking	7
Badminton (social)	4.5
Baking	4
Ballet	6
Ballroom dancing	5
Bartending	2.5
Baseball (general)	5
Basketball	7
Biking (light)	6
Biking (moderate)	8
Biking (racing)	16
Biking (vigorous)	10
Biking, stationary (moderate)	7
Board surfing	3

Activity list	Metabolic rate factor
Bobsledding	7
Body work (automobile)	4.5
Bowling	3
Boxing (general)	12
Broomball	7
Bungee jumping	4
Calisthenics	6
Canoeing (carrying)	7
Canoeing (moderate)	7
Carpentry	3
Carrying (heavy weight)	10
Carrying (moderate weight)	8
Carrying golf clubs	5.5
Carrying logs	11
Carrying wood	5
Caulking	4.5
Changing light bulb	2.5
Changing linen	2.5
Child care	3
Chopping wood (light)	5
Chopping wood (vigorous)	17
Cleaning the garage	4.5
Clearing dishes from table	2.3
Clearing land	5
Climbing hills (light backpack)	7
Climbing hills (moderately heavy backpack)	8
Climbing ladder	8
Coaching sports	4
Computer work (typing, sitting)	1.5
Concrete work	7
Conducting music	2.5
Construction work	5.5
Cooking	2.5
Cricket	5
Croquet	2.5

Activity list	Metabolic rate factor
Cross-country hiking	6
Cross-country running	9
Cross-country skiing (light)	7
Cross-country skiing (moderate)	8
Cross-country skiing (vigorous)	9
Curling	4
Cutting trees	8
Dancing	4.5
Darts	2.5
Digging ditches	8.5
Directing traffic	2.5
Disco dancing	5.5
Diving	3
Dressing/undressing	2.5
Driving	3
Driving range (golf)	3
Driving truck, including loading/unloading	6.5
Dusting	2.5
Eating	1.5
Electrical work	3
Feeding farm animals	4
Fertilizing lawn	2.5
Field hockey	8
Finishing/refinishing furniture	4.5
Fishing	4
Folding and hanging clothes	2
Folk dancing	5.5
Football	8.5
Frisbee	3
Gambling (casino)	2
Gardening	5
General household cleaning	3.5
Golf (general)	4.5
Golf (with cart)	3.5
Grocery shopping	3.5

Activity list	Metabolic rate factor
Gymnastics	4
Hair Dressing	2.5
Handball	8
Hand tool work	1.5
Hang gliding	3.5
Hanging sheet rock	4.5
Hanging wallpaper	4.5
Health club exercise (general)	5.5
Hoeing	5
Horseback riding (galloping)	8
Horseback riding (trotting)	6.5
Horseback riding (walking)	2.5
Horseshoes (game)	3
Hunting (general)	5
Ice fishing	3
Ice hockey	8
Ice/inline skating (light)	5.5
Ice/inline skating (moderate)	9
Ice/inline skating (vigorous)	13
Installing or removing carpet	4.5
Installing rain gutters	6
Installing tile or linoleum	4.5
Ironing	2.3
Jackhammer work	6
Jogging	7
Kick boxing	10
Kickball	7
Knitting	1.5
Lab work	1.5
Lacrosse	8
Laundry	2
Lawn bowling	3
Listening to music	1
Loading/unloading wood or lumber	5
Locksmith work	3.5

Activity list	Metabolic rate factor
Lying (still)	0.9
Making beds	2
Marching (rapid speed)	6.5
Marching (regular speed)	3.5
Martial arts	10
Massage work	4
Mechanical repair	3
Miniature golf	3
Mopping	4.5
Motocross	4
Mountain biking	8.5
Moving	7
Mowing lawn	4.5
Nursing (patient care)	3
Office work (general)	2.5
Officiating (running)	7
Officiating (standing)	4
Operating lathe	3
Operating punch press	5
Operating snow blower	4
Packing/unpacking	3
Paddleball (light)	6
Paddleball (vigorous)	10
Paddle boating	4
Painting	4.5
PC/computer repair	2
Personal care	2.5
Ping pong/table tennis	4
Pistol shooting	2.5
Planting trees	4.5
Plastering	4.5
Playing cello	2
Playing board games	1.5
Playing cards	1.5
Playing drums	4

Activity list	Metabolic rate factor
Playing flute	2
Playing guitar	3
Playing piano	2.5
Playing trumpet	2.5
Playing violin	2.5
Plumbing	3.5
Polo (water)	10
Poultry work	8
Power boating	2.5
Pressing (tailoring)	4
Printing work	2.3
Pull-ups/Push-ups/Sit-ups	8
Pushing baby stroller	2.5
Putting away groceries	2.5
Race walking	6.5
Racquetball (competitive)	10
Racquetball (leisure)	7
Raking lawn	4
Reading	1.3
Repair Work (heavy parts - standing)	3
Repair Work (light parts - sitting)	1.5
Riding in a Car or other Vehicle	1
Rock climbing	11
Roofing	6
Rugby	10
Running (fast)	18
Running (moderate)	12.5
Running (slow)	8
Running (up stairs)	15
Sailing	4
Sawing hardwood, by hand	7.5
Sawing wood by hand	7
Sawing wood with power saw	4.5
Scraping paint	4.5
Scrubbing floors (on hands and knees)	5.5

Activity list	Metabolic rate factor
Seeding lawn	2.5
Serving food	2.5
Setting table	2.5
Sewing	2
Sexual activity	1.3
Shoe repair	2.5
Shopping	2.3
Shoveling	7
Sitting	1
Sit-ups	8
Skateboarding	5
Ski machine	9.5
Skiing	6
Skin diving (snorkeling only)	5
Skin/scuba diving (fast)	16
Skin/scuba diving (light)	7
Skin/scuba diving (moderate)	12.5
Sky diving	3.5
Sledding	7
Sleeping	0.9
Snow shoeing	8
Snowboarding	6
Snowmobiling	3.5
Soccer (competitive)	10
Soccer (leisure)	7
Softball	5
Splitting logs	6
Spreading dirt	5
Squash	12
Standing	1.2
Stationary rowing (moderate)	7
Step aerobics	8
Stocking shelves	3
Store clerk work	2.5
Stretching	4

Activity list	Metabolic rate factor
Studying	1.8
Surfing the Internet or TV	1.5
Sweeping	3
Swimming (light)	6
Swimming backstroke	8
Swimming breaststroke	10
Swimming butterfly	11
Swimming crawl/freestyle	10
Synchronized swimming	8
Tai chi	4
Taking a bath	2
Taking a shower	4
Talking	1.5
Tennis	7
Tobogganing	7
Trampoline	3.5
Trap shooting	2.5
Treadmill, 10% incline (fast – 10 mph)	24.24
Treadmill, 10% incline (moderate – 6 mph)	14.98
Treadmill, 10% incline (slow – 2 mph)	5.5
Treadmill, 15% incline (fast – 10 mph)	27.82
Treadmill, 15% incline (moderate – 6 mph)	17.14
Treadmill, 15% incline (slow – 2 mph)	6.92
Treadmill, 5% incline (fast – 10 mph)	20.65
Treadmill, 5% incline (moderate – 6 mph)	12.82
Treadmill, 5% incline (slow – 2 mph)	4.07
Treadmill, level (fast – 10 mph)	17.06
Treadmill, level (moderate – 6 mph)	10.66
Treadmill, level (slow – 2 mph)	2.65
Trimming shrubs	4
Trimming trees	9
Vacuuming	2.5
Volleyball (beach)	8
Volleyball (competitive)	4
Volleyball (leisure)	3

Activity list	Metabolic rate factor
Volleyball (water)	3
Walking	2
Walking (down stairs)	3
Walking (inside home)	3
Walking (moderate – 4 mph)	4
Walking (race walking)	6.5
Walking (up stairs)	8
Walking downhill	3
Walking uphill	6
Washing dishes	2.3
Washing fence/windows	4.5
Washing/waxing car or boat	4.5
Watch repair	1.5
Watching TV or a movie	1
Water aerobics	4
Water skiing	6
Watering lawn or garden	1.5
Weeding	4
Weight lifting, moderate	3
Weight lifting, vigorous	6
Welding	3
Wheeling self in wheelchair	3.5
Whirlpool (sitting)	1
Whitewater canoeing/kayaking/rafting	5
Windsurfing	3
Wrapping presents	1.5
Wrestling	6
Writing	1.8
Yoga	4

Sample food logs

Shaping a New You			
Daily written record of energy intake			
Date: ______ Morning weight________ Hunger level H M L			
Glasses of water: 1 2 3 4 5 6 7 8			
Time	Food	Portion Size	Calories
8:15 am	Lender's blueberry bagel	1-(3.6 oz)	280
	Light cream cheese	2 oz	120
	Coffee (regular)	2 cups	10
	Half and half	2 T.	40
	Sugar	2 packets	50
10:30 am	Regular soda	12oz	150
1:15 pm	Lois Rich turkey	3 slices	50
	Lettuce	2 leaves	6
	Bread (white, thin sliced)	2 slices	140
	Mayonnaise (light)	1 T.	50
	Potato chips	4 oz bag	600
	Cookies (chocolate chip)	2 small	160
6:30pm	Pasta with marinara sauce	2 cups	530
	French bread	2 pieces	220
	Red wine	2 glasses (8 oz)	160
	Butter	1 T.	100
9:30pm	Haagen Dazs- Mint chip	1 cup	600
TOTAL CALORIES			3,266

The food diary can easily be modified to go with your chosen weight loss program. For example, if you also want to count fat grams, then a column for fat grams can be added. An example of this log is provided on the next page.

Shaping a New You

Daily written record of energy intake

Date: _______ Morning weight_________ Hunger level H M L

Glasses of water: 1 2 3 4 5 6 7 8

Time	Food	Portion Size	Protein (g)	Carbs (g)	Fat (g)	Calories
Totals						

Weekly weight management record

Shaping a New You

Starting weight: ______ Future weight goal: ______

Height: ____"(in inches) Starting waist measurement ____inches

Starting BMI: ________ Starting RMR:________

Weight management goals

A. Weight loss goal: ______pounds (current minus future weight goal)

B. Time frame: ______weeks

C. Weight loss goal: (A)______ ÷ (B) ______= _______ pound(s) per week

D. Energy Balance goal: _____ pound(s) per week X 3,500 calories ÷ 7 =______calories per day of negative Energy Balance

Week	Weight	Waist		Week	Weight:	Waist:
1				27		
2				28		
3				29		
4				30		
5				31		
6				32		
7				33		
8				34		
9				35		
10				36		
11				37		
12				38		
13				39		
14				40		
15				41		
16				42		
17				43		
18				44		
19				45		
20				46		
21				47		
22				48		
23				49		
24				50		
25				51		
26				52		

TIPS FOR LONG-TERM WEIGHT MANAGEMENT

- Know your energy intake.
    - Record everything you eat and drink.
    - Read labels carefully and measure portions.
- Minimize overeating.
    - Eat regular meals and snacks to prevent hunger.
    - Don't skip meals.
    - Keep your home stocked with healthy low calorie foods.
- Monitor your weight.
    - Weigh yourself daily or weekly.
    - Learn your normal weight fluctuations.
    - Act when your weight increases more than 5 pounds.
- Set realistic goals when trying to lose weight.
    - A 10% weight loss is a good place to start.
    - 1–2 pounds of weight loss a week is realistic and healthy.
- Make a commitment to increase your physical activity.
    - Build your activity time into your schedule.
    - Consider your activity appointment with yourself to be mandatory.
    - Accumulate activity throughout your day.

SMART TIPS ON PORTION SIZE

- 1 cup of cereal, pasta or vegetables = 1 fist
- 1 baked potato = 1 fist
- 1 ounce of cheese = 1 finger
- 1 teaspoon of butter, peanut butter = the tip of a thumb
- 3 ounces of meat = 1 palm
- 1 ounce of chips or nuts = 1 handful
- A typical bakery bagel is 4 ounces.
- A pasta side dish at an Italian restaurant is typically 2 cups.
- A steak at a typical restaurant is 8–10 ounces.
- A large "Big Grab" sack of chips is 4 ounces.
- A regular size can of soda is 12 ounces.

HOW TO READ A FOOD LABEL

Whole Milk
Serving Size 8 fl oz (240mL)
Servings Per Container 2

Amount Per Serving

Calories 150 Calories from Fat 70

	% Daily Value*
Total Fat 8g	**12%**
Saturated Fat 5g	**25%**
Cholesterol 35mg	**12%**
Sodium 125mg	**5%**
Total Carbohydrates 12g	**4%**
Dietary Fiber 0g	**0%**
Sugars 11g	
Protein 8g	

Vitamin A 6% • Vitamin C 14%

Calcium 30% • Iron 0% • Vitamin D 25%

*Percent Daily Values are based on a 2,000 calorie diet. Your daily values may be higher or lower depending on tour calorie needs

	Calories	2,000	2,500
Total Fat	Less than	65 g	80 g
Sat Fat	Less than	20 g	25 g
Cholesterol	Less than	300 mg	300 mg
Sodium	Less than	2,400 mg	2,400 mg
Total Carbohydrates		300 g	375 g
Dietary Fiber		25 g	30 g

Serving Size: All label information refers to a single serving. In the example, if you have two servings (16 oz), multiply all the information on the label by two.

Calories: Total amount of food energy present in one serving.

Calories from Fat: Total number of calories that come from fat -- one gram of fat = nine calories. Multiply the number of grams of fat by nine to get total fat calories.

Total Fat: Total amount of fat grams per serving.

% **Daily Value from Fat:** This percentage of fat is based on a 2000 calorie diet. Foods that are less than 5% are usually low-fat choices.

Saturated Fat: Amount of saturated fat (cholesterol-raising fat) within the total fat grams.

Cholesterol: Mostly found in animal products. Try to keep intake to less than 300 mg/day.

Sodium: Implicated in high blood pressure. Keep intake below 2,400 mg/day. Any product over 450 mg is considered a high-sodium food.

Total Carbohydrate: Total amount of carbohydrate grams per serving. One gram of carbohydrate equals four calories. Multiply number of carbohydrate grams by four to get calories from carbohydrate.

Protein: Total amount of protein grams per serving. one gram of protein equals four calories. Multiply number of protein grams by four to get protein calories.

Appendix A

Reference Library

APPENDIX A. REFERENCE LIBRARY

TABLE OF CONTENTS

FIGURES AND TABLES **172**
COMPONENTS OF WEIGHT LOSS **174**
WATER **183**
CARBOHYDRATES **185**
FATS AND FATTY ACIDS **190**
FIBER **198**
PROTEIN AND AMINO ACIDS **204**
VITAMINS **209**
BIOTIN (VITAMIN H) 209
CHOLINE 212
FOLIC ACID (FOLATE) 214
PANOTHENTIC ACID (VITAMIN B5) 218
VITAMIN A 222
VITAMIN B1 (THIAMIN) 225
VITAMIN B2 (RIBOFLAVIN) 227
VITAMIN B3 (NIACINAMIDE OR NIACIN) 230
VITAMIN B6 (PYRIDOXINE) 232
VITAMIN B12 (COBALAMIN) 236
VITAMIN C (ASCORBIC ACID) 239
VITAMIN D 245
VITAMIN E 247
VITAMIN K 250
MINERALS **254**
CALCIUM 254
CHROMIUM 258
COPPER 260
FLUORIDE 262
IRON 264
MAGNESIUM 266
MANGANESE 268
PHOSPHORUS 270
POTASSIUM 273
SELENIUM 274
ZINC 277

WEIGHT-RELATED ILLNESSES **279**
DIABETES 279
SLEEP APNEA 281
GOUT 283
THE METABOLIC SYNDROME 284
CORONARY HEART DISEASE AND CHOLESTEROL 286
SPECIAL TOPICS **290**
HOMOCYSTEINE AND CHOLESTEROL 290
SWEETENERS 293
FAT SUBSTITUTES 295
APPENDIX B. WHAT IS COLORADO ON THE MOVE™? **301**
STEPS MAKE MODERATE EXERCISE EASY 302

FIGURES AND TABLES

CALCULATIONS FOR BODY MASS INDEX **175**
WAIST CIRCUMFERENCE MEASUREMENT **176**
MODERATE PHYSICAL ACTIVITIES **180**
PHYSICAL ACTIVITIES **181**
SOURCES OF FIBER **201**
PROTEIN REQUIREMENTS **205**
FOOD SOURCES OF PROTEIN **207**
BIOTIN INTAKE LEVELS BY AGE **210**
FOOD SOURCES OF BIOTIN **211**
SOURCES OF CHOLINE **213**
FOOD SOURCES OF FOLIC ACID **216**
PANTOTHENIC ACID REQUIREMENTS **219**
FOOD SOURCES OF PANTOTHENIC ACID **220**
VITAMIN A REQUIREMENTS **222**
SOURCES OF VITAMIN A **224**
RDA FOR VITAMIN B2 ..... **229**
RDA FOR VITAMIN B3 ..... **231**
FOOD SOURCES FOR VITAMIN B6 ..... **234**
RDA FOR VITAMIN B12 ..... **238**
SOURCES OF VITAMIN C **241**
ADEQUATE INTAKES FOR VITAMIN D **246**
RDA FOR VITAMIN E **248**
FOOD SOURCES FOR VITAMIN E **249**

FOOD SOURCES OF VITAMIN K 252
BASIC CALCIUM REQUIREMENTS 255
FOOD SOURCES OF CALCIUM 256
FOOD SOURCES OF CHROMIUM 259
FLUORIDE REQUIREMENTS 262
FOOD SOURCES OF FLUORIDE 263
FOOD SOURCES OF IRON 265
RDAS FOR MAGNESIUM 266
FOOD SOURCES OF MAGNESIUM 267
MANGANESE REQUIREMENTS 269
RDA FOR PHOSPHORUS 271
FOOD SOURCES OF PHOSPHORUS 271
SELENIUM REQUIREMENTS 275
ZINC REQUIREMENTS 278
SUMMARY OF POSSIBLE LIFESTYLE CHANGES 287
CLASSIFICATION OF FAT SUBSTITUTES BY NUTRIENT SOURCE, FUNCTIONAL PROPERTIES AND USE IN FOOD 296

As we have seen in Parts I and II of this book, weight and weight loss are complex issues – issues that are confronted by millions of North Americans daily. The obesity rate is growing steadily, partially because of sedentary lifestyles and an all-too-typical North American diet that is high in fats, refined carbohydrates, and calories. The need for weight loss products has become a major consumer concern.

We've included this section so that you will have a reference library at your fingertips.

COMPONENTS OF WEIGHT LOSS

Theoretically, weight loss is a simple concept that can be summarized as input vs. output. Essentially this concept refers to the amount of food or calories ingested vis-à-vis the amount of energy expended.

The body requires a certain number of calories or energy to carry on everyday activities; this is referred to as **total energy expenditure**. Total energy expenditure is comprised of three components:

- the resting metabolic rate (RMR),
- the thermic effect of food, and
- the thermic effect of exercise.

The **resting metabolic rate** is the total energy required by the body in its resting state. The body requires this basal amount of energy just to sustain life processes such as breathing and involuntary physiological functions.

The **thermic effect of food** is the percentage of your total energy expenditure that is used to digest food.

The **thermic effect of exercise** refers to the amount of energy you use to carry out physical activity.

In weight loss or weight management, several parameters are important. When referring to weight issues, the terms **body mass index** or BMI and Waist Circumference are often used as indices.

Body mass index is a formula that is used for determining obesity and as an assessment for relative risk of developing obesity associated diseases such as cardiovascular disease, diabetes and dyslipidemias. BMI is used as a surrogate measure of total body fat. It is calculated by dividing a person's weight in kilograms by the square of the person's height in meters. If calculating from pounds, the weight is multiplied by 703 and divided by the height (in inches) squared.

Calculations for Body Mass Index

BMI	19	20	21	22	23	24	25	26	27	28	29	30	35	40
Height (in.)	Weight (lb.)													
58	91	96	100	105	110	115	119	124	129	134	138	143	167	191
59	94	99	104	109	114	119	124	128	133	138	143	148	173	198
60	97	102	107	112	118	123	128	133	138	143	148	153	179	204
61	100	106	111	116	122	127	132	137	143	148	153	158	185	211
62	104	109	115	120	126	131	136	142	147	153	158	164	191	218
63	107	113	118	124	130	135	141	146	152	158	163	169	197	225
64	110	116	122	128	134	140	145	151	157	163	169	174	204	232
65	114	120	126	132	138	144	150	156	162	168	174	180	210	240
66	118	124	130	136	142	148	155	161	167	173	179	186	216	247
67	121	127	134	140	146	153	159	166	172	178	185	191	223	255
68	125	131	138	144	151	158	164	171	177	184	190	197	230	262
69	128	135	142	149	155	162	169	176	182	189	196	203	236	270
70	132	139	146	153	160	167	174	181	188	195	202	207	243	278
71	136	143	150	157	165	172	179	186	193	200	208	215	250	286
72	140	147	154	162	169	177	184	191	199	206	213	221	258	294
73	144	151	159	166	174	182	189	197	204	212	219	227	265	302
74	148	155	163	171	179	186	194	202	210	218	225	233	272	311
75	152	160	168	176	184	192	200	208	216	224	232	240	279	319
76	156	164	172	180	189	197	205	213	221	230	238	246	287	328

For men and women, a BMI of 25 to 29.9 kg/m^2 is considered overweight while a BMI greater than 30 kg/m^2 is considered obese. A BMI of 30 indicates that the individual is about 30 pounds overweight.

It is, however, important to note that the BMI is an ineffective measure where there is extreme muscularity, as in performance athletes, the presence of edema, muscle wasting, and in individuals of short stature.

In addition, the relationship between BMI and total body fat may vary with age, gender, and possibly ethnicity. However, BMI used in conjunction with waist circumference, is generally a good indicator of obesity and relative risk of developing certain obesity related chronic diseases.

Waist circumference is a measure of abdominal fat deposits. It is generally considered safer to deposit fat subcutaneously around the hips and thighs than around the internal organs in the abdominal region. This is because fat deposited around the visceral organs is more metabolically active and affects metabolism in ways that contribute to increased disease risk.

Waist Circumference Measurement[5]

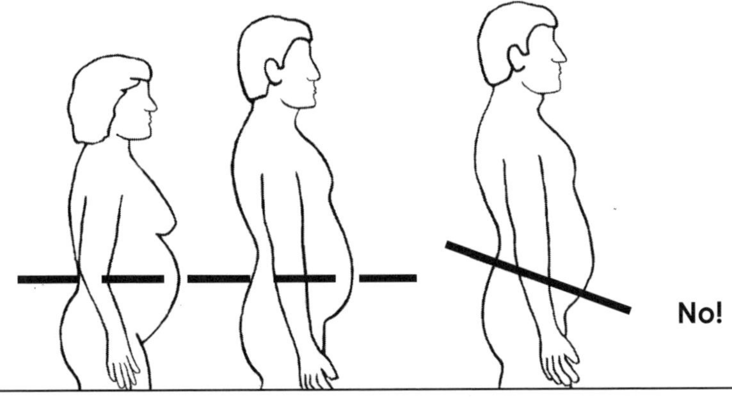

To measure waist circumference, locate the upper hip bone and the top of the right iliac crest. Place a measuring tape in a horizontal plane around the abdomen at the level of the iliac crest. The tape should be just below the bellybutton. Waist circumference is often not the same as a person's pant waist size. Before reading the tape measure ensure that the tape is snug, but does not compress the skin, and is parallel to the floor. The measurement is made at the end of a normal breath just before inhaling again.

WEIGHT LOSS THROUGH PHYSICAL ACTIVITY AND CALORIC RESTRICTION

One pound of body weight is equal to about 3,500 kilocalories (commonly referred to as calories). Thus to lose one pound of weight, energy intake must be restricted by 3,500 calories. With a caloric restriction between 500 and 1,000 calories per week, it is reasonable to expect a one to two pound weight loss per week and a 10% reduction in body weight in six months.

This weight loss is achieved by decreasing the number of calories ingested and increasing the amount of exercise or activity. Weight loss achieved only through caloric restriction is not recommended because, in addition to the loss of fat, this method causes loss of **fat-free mass**. A decrease in fat-free mass will result in a decreased resting metabolic rate, slowing the rate of weight loss.

Exercise must be incorporated into the weight loss regime to achieve long-term weight control. Increasing the amount of exercise will not only burn more calories, it will also increase the resting metabolic rate and play a major role in the prevention of weight regain[5]. The faster metabolism causes the body to use energy more efficiently. In addition, increased exercise will have added benefits—it will increase alertness, energy, and concentration. Most importantly, it will decrease the risk of developing coronary heart disease, hypertension, non-insulin dependent diabetes mellitus, and colon cancer.

There is a common misconception that weight loss involves only the amount of weight lost and the time it takes to lose the weight. However, weight maintenance is an extremely important part of an effective weight loss program. In fact, weight maintenance is crucial.

Weight maintenance refers to maintaining the desired weight after the diet and exercise regime is complete and the target weight has been achieved. Successful weight maintenance is defined as regaining no more than 6.6 pounds (3 kilograms) within two years and maintaining a reduction in waist circumference of at least 1.6 inches or 4 cm[5].

Weight maintenance is the more difficult component of weight loss because there is no rewarding endpoint (e.g., decreasing from a size 14 to a size 12). Weight maintenance is achieved through continuous dietary vigilance and a regular program of physical activity.

BENEFITS OF PHYSICAL ACTIVITY

As stated before, increased physical activity brings many health benefits. As summarized by Haskell, in observational studies individuals who were more physically fit had lower rates of non-insulin dependent diabetes mellitus. In addition, people at high risk who incorporated moderate changes in physical activity levels, diet (decreased intake of saturated fat), and body weight lowered their risk of developing non-insulin dependent diabetes mellitus[1].

In regard to cardiovascular disease, previously sedentary individuals who began to expend 1,500 calories a week or more in vigorous activity substantially lowered their cardiovascular risk[1]. Researchers have also found that individuals suffering from depression showed excellent improvement in mood and energy levels when they began engaging in physical activity[3]. Researchers have also concluded that 20% of premature mortality (early death) could have been avoided through daily physical activity[4].

HOW MUCH PHYSICAL ACTIVITY IS REQUIRED?

It is a common misconception that to lose weight one must exercise vigorously enough to sweat. It's not true. Any physical activity that increases heart rate is valid exercise.

However, the time it takes to burn a given amount of calories depends on the intensity of the physical activity. Experts tell us that just 30 to 60 minutes of physical activity each day is enough to improve health. (Canadian Guidelines recommend 60 minutes of daily activity, while US Guidelines recommend 30 to 60 minutes.)

Although 60 minutes seems like a lot, it can be broken up into shorter periods of activity. For example, a 10 minute walk in the morning, a 20 minute walk at noon, 10 minutes of stairs during a coffee break, and a 20 minute walk after dinner. The key to success is choosing activities that you enjoy and pacing yourself.

At the Whistler 2001 Communicating Physical Activity and Health Messages[19] seminar it was pointed out that, in technologically advanced societies (for example, North American society), adults who are normally sedentary should add close to 60 minutes of moderate to vigorous physical activity to their daily routines to prevent unhealthy weight gain.

WHAT TYPE OF ACTIVITY SHOULD I DO?

According to Canada's Physical Activity Guide[2], physical activity can be divided into three types:

- endurance activities,
- flexibility activities, and
- strength activities.

Endurance activities are activities that increase energy levels and help the heart, lungs, and circulatory system stay healthy. Examples of endurance exercises are gardening, walking, skating, and household chores. These activities should be eased into, beginning with short periods at low intensity and progressing to longer, more challenging periods.

Flexibility activities help the body move more easily and fluidly, by keeping muscles relaxed and joints mobile. This improves the quality of life and helps maintain independence as we get older. Flexibility exercises include Tai Chi, Yoga, gardening, and vacuuming.

Strength activities maintain bone and muscle strength and prevent diseases such as osteoporosis. Strength activities are exercises that work muscles against some form of resistance. Strengthening activities include raking leaves, lifting and carrying groceries or children, and climbing stairs[2].

---

[19] The Whistler 2001 Communicating Physical Activity and Health Messages seminars were hosted by Health Canada and The United States Centers for Disease control and Prevention and organized by ParticipACTION. The aim of the seminars was to encourage collaboration among public health, private and voluntary organizations, industry, and the media to develop consensus strategies to promote and encourage physical activity.

For a physically inactive individual, strengthening activities should involve two sessions of resistance exercise each week. A single session should include 8 to 10 different exercises, with one or two sets of each exercise performed and 10 to 12 repetitions per set. At the end of each exercise, effort should be near maximum.

Moderate Physical Activities[5]

Examples of moderate amounts of physical activity* *These start with the least vigorous in each category and increase in intensity as you progress down the list***	
Common Chores	**Sporting Activity**
Washing and waxing a car for 45-60 minutes	Playing Volleyball for 45-60 minutes
Washing windows or floors for 45-60 minutes	Playing touch football for 45 minutes
Gardening for 30-45 minutes	Walking 1¾ miles in 35 minutes (20 minutes per mile)
Pushing stroller 1 ½ miles in 30 minutes	Basketball (shooting baskets) for 30 minutes
Raking leaves for 30 minutes	Bicycling 5 miles in 30 minutes
Walking 2 miles in 30 minutes	Dancing fast (social) for 30 minutes
Shoveling snow for 15 minutes	Water aerobics for 30 minutes
Stair walking for 15 minutes	Basketball (playing a game for 15-20 minutes)
	Jumping rope for 15 minutes
	Running 1½ miles in 15 minutes (15 minutes per mile)
*A moderate amount of physical activity is roughly equivalent to one that uses about 150 calories of energy per day or about 1000 per week.	
**Some activities can be performed at various levels of intensity. The suggested durations correspond to the expected level of intensity.	

Physical activities	
Type	**Activity**
Very light activity	Increased standing, room painting, pushing a wheelchair, yard work, ironing, cooking, playing musical instrument.
Light activity	Slow walking (24 min/mile), garage work, carpentry, house cleaning, child care, golf, sailing, table tennis.
Moderate activity	Walking (15 min/mile), weeding and hoeing garden, carrying a load, cycling, skiing, tennis, dancing.
High activity	Jogging a mile in 10 minutes, walking uphill with a load, tree felling, heavy manual digging, basketball, climbing, soccer.
Other key activities	Flexibility exercises to attain full range of joint motion, strength or resistance exercises, aerobic conditioning.

REFERENCES

1. Haskell W. The physical activity profile required to produce health benefits. Effects of type, intensity, frequency and duration of activity on health outcomes. *Whistler 2001 Communicating Physical Activity and Health Messages: Science into Practice.* Health Canada and Centers for Disease Control.

2. Health Canada. *Canada's Physical Activity Guide to Healthy Active Living.*

3. Rejeski, W. J. & Brawley, L. R. Psychological outcomes form physical activity: Implications for messaging. *Whistler 2001 Communicating Physical Activity and Health Messages: Science into Practice.* Health Canada and Centers for Disease Control.

4. Lee, I. M. Population attributable risk of no physical activity. *Whistler 2001 Communicating Physical Activity and Health Messages: Science into Practice.* Health Canada and Centers for Disease Control.

5. National Institutes of Health; National Heart, Lung and Blood Institute and North American Association for the Study of Obesity. *The Practical Guide: Identification, Evaluation and Treatment of Overweight and Obesity in Adults.* October 2000. NIH publication No. 00-4084 . Available in PDF format through the Internet http://www.nhlbi.nih.gov/guidelines/obesity/practgde.htm

WATER

Water is a chemical compound comprised of two molecules of hydrogen and one molecule of oxygen. Combined, they have the chemical formula H_2O.

Water is essential to every living organism. The human body is more than half water. The majority of this water is in the cells, while the remainder is in the extracellular compartments and in the blood.

What does water do?

Water is involved in many of the body's integral processes: it is involved in the transport of blood components, is a medium for dissolving and transporting nutrients, is a medium for many intracellular reactions, and acts as a major vehicle for the disposal of bodily waste[2]. Water also plays a major role in regulating body temperature through perspiration.

Humans can sweat up to ten liters a day[1]. Dehydration or hypohydration (too much water) can dramatically decrease muscular endurance and physical work capacity[1].

DEHYDRATION

When we don't have enough water in our bodies, we become dehydrated. In humans, symptoms of dehydration may begin at a water deficiency as low as 1%. Symptoms of dehydration include thirst, dizziness, weakness, tiredness, inability to concentrate, headache, and nausea. Death is the final outcome of continued dehydration[1].

SOURCES OF WATER

The most obvious source of water in the diet is plain water itself. It is a component of other beverages such as milk or juice. Another important and often discounted source of water is food. Food, especially fruits and vegetables, contains lots of water.

It is important to note that certain beverages actually decrease the amount of water in the body. Beverages that contain caffeine act as diuretics, causing the body to excrete water in the form of urine.

REFERENCES

1. Askew, E. W. Water. In E. E. Ziegler & L.J. Filer (Eds.) *Present Knowledge in Nutrition,* 7th Edition. Washington DC: ILSI Press, 1996: pp. 98-108.

2. Kiester, E. (Ed.). Nutrition. In *New Family Medical Guide*. Des Moines, Iowa: Meredith Corporation, 1989: pp. 250-263.

CARBOHYDRATES

The body is a complicated machine with many parts and it needs fuel to function. The body's main fuel source is carbohydrate. One gram of carbohydrate is equal to 4 calories of energy.

---

1 gram of carbohydrate = 4 calories

Carbohydrates provide the body with the energy to carry out everyday activities. Carbohydrates make up the majority of living matter on earth, so it's easy to understand why experts recommend that approximately 55% of what we eat should be in the form of carbohydrates.

WHAT ARE CARBOHYDRATES?

Carbohydrates are compounds of carbon, hydrogen, and oxygen created by plants during photosynthesis. **Primary carbohydrates** are known as simple sugars, or saccharides. When these sugars attach to each other to form long branching chains, they are known as **complex carbohydrates** (fiber), or starches.

WHAT DO CARBOHYDRATES DO?

Carbohydrates are an efficient source of energy for the body. The body also uses carbohydrates to produce structural and functional materials.

When complex carbohydrates are eaten, the digestion process changes them into simple sugars that are quickly absorbed into the blood and carried to tissues in the body. These sugars are then used to create energy through metabolic processing.

The body uses two main forms of carbohydrates: glucose and glycogen.

- **Glucose** is the major form used for the body's immediate energy needs. Surplus glucose is stored as glycogen.
- **Glycogen** is a chain of glucose and is stored in liver and muscles. Excess glucose may be converted into storage fat; however, this is rare unless carbohydrates have been overeaten for many consecutive days. Gaining weight on a high-carbohydrate diet usual occurs because the ingested carbohydrate suppresses the body's burning of fat. As a result, any fat eaten along with the carbohydrate is stored and fat stores swell.

Carbohydrates also serve many other functions in addition to supplying the body with energy. For example, **dietary fiber** is a class of carbohydrate that the body is unable to digest. Some fibers ferment in the colon, providing energy for intestinal cells while other fibers travel the entire length of the bowel undigested and unmetabolized. It is believed that dietary fiber protects against colon cancer and improves overall bowel health.

Recently, attention has been focused on the glycemic response to dietary carbohydrates. Glycemic response is the change in blood glucose that occurs following ingestion of food. Many simple starches and sugars are rapidly absorbed, leading to a high glycemic response. A **high glycemic response** is thought to increase appetite and promote chronic diseases related to insulin, such as diabetes.

Complex carbohydrates such as those from whole grains have **low glycemic response** and are among the carbohydrates most recommended by health authorities. Although many fruits are composed of more than 90% carbohydrates, the carbohydrates in fruit are primarily simple sugars like glucose and fructose. The best dietary sources of complex carbohydrates are rice, corn, whole grain flours and pastas, potatoes, yams, beans, and peas.

Much attention has also been focused on the use of **probiotics** to promote intestinal health and general wellness. Certain carbohydrates that are undigested by the gastrointestinal tract have been found to promote the growth of these beneficial bacterial in the intestinal tract. This could have many implications for individuals suffering from Crohn's disease or colitis[4,5,6] or colon cancer[7,8,9], or more generally on intestinal health[10].

Finally, besides being good sources of energy, foods high in carbohydrates—especially those that are not highly processed—are also generally rich in vitamins and minerals that are important for good health.

SOURCES OF CARBOHYDRATES

Simple and complex carbohydrates are found in the stems, leaves, roots, and fruits of plants. Animal products other than milk, which contains lactose, do not contain significant amounts of carbohydrates. Lactose is the primary source of carbohydrates for babies.

CARBOHYDRATE REQUIREMENTS

According to the USDA recommended Acceptable Macronutrient Distribution Range, we should eat between 45% and 65% carbohydrates, in the form of complex carbohydrates or natural sugars like those found in fruits. Thus, in a 2,400 calorie diet, 1,200 to 1,440 calories of the diet should be carbohydrates.

Carbohydrates Should Be
55% of Total Calorie Intake

For additional information go to www.nap.edu and search for dietary reference intakes for energy, carbohydrates, fiber, fat, protein, and amino acids.

REFERENCES

1. Pick, M. E., Hawrysh, Z. J. , Gee, M. I. & Toth. E. Barley bread products improve glycemic control of type 2 subjects. *Int J Food Sci Nutr.* 1998; 49(1):71-78.

2. Urooj, A., Vinutha, S. R., Puttaraj, S., Leelavathy, K. & Haridasa-Rao, P. Effect of barley incorporation in bread on its quality and glycemic responses in diabetes. *Int J Food Sci Nutr.* 1998; 49(4): 265-270.

3. Yokoyama, W. H., Hudson, C.A., Knuckles, B. E., Chiu, M. C. M., Sayre, R. N., Turnlund, J. R. & Schneeman, B. O. Effect of barley beta-glucan in durum wheat pasta on human glycemic response. *Cereal Chem.* 1997; 74(3): 293-296.

4. Madsen, K. L. Inflammatory bowel disease: Lessons from the IL-10 gene deficient mouse. *Clin Invest Med.* 2001;24(5): 250-257.

5. Guslandi, M. Probiotics in Crohn's disease. *Gut.* 49(6): 873-873.

6. Oviedo, J. & Farraye, F. A. Self-care for the inflammatory bowel disease patient: what can the professional recommend? *Sem In Gastro Dis.* 2001;12(4):223-236.

7. Wollowski, I., Rechkemmer, G. & Pool-Zobel, B. L. Protective Role of Probiotics and prebiotics in colon cancer. *Am J Clin Nutr.* 2001; 73(2S): 451S-455S.

8. Brady, L. J., Gallaher, D. D. & Busta, F .F. The role of probiotic cultures in the prevention of colon cancer. *J nutr.* 2000; 130(2S):410S-414S.

9. Sanders, M. E., Klaenhammer, T. R. Invited review: the scientific basis of *Lactobacillus acidophilus* NCFM functionality as a probiotic. *J dairy sci.* 2001:84(2): 319-331.

10. Marteau, P.R., de Vrese, M,. Cellier, C. J. & Schrezenmeir, J. Protection from gastrointestinal disease with the use of probiotics. *Am J Clin Nutr. 2001; 73(2S): 430S-436S.*

11. Ludwig, D. S. The Glycemic Index. Physiological mechanisms relating to obesity, diabetes and cardiovascular disease. *JAMA*. 2002;287(18):2414-2423.

12. Jenkins, D. J. A., Wolever, T. M. S., Taylor, R. H., Barker, H. M., Fielden, H., Baldwin, J. M., Bowling, A. C., Newman, H. C., Jenkins, A. L. & Goff, D. V. Glycemic index of foods: a physiological basis for carbohydrate exchange. *American Journal of Clinical Nutrition*. 1981;34:362-366.

13. Wolever, T. M. S. & Jenkins, D. J. A. The use of the glycemic index in predicting the blood glucose response to mixed meals. *American Journal of Clinical Nutrition*. 1986;43:167-172.

14. Ha, M. A., Mann, J. I., Melton, L. D. & Lewis Barned, N. J. Calculation of the glycaemic index. *Diabetes, Nutrition and Metabolism, Clinical and Experimental*. 1992; 5:137-139.

15. Institute of Medicine. Food and Nutrition Board. *Dietary Reference Intakes for Energy, Carbohydrates, Fiber, Fat, Protein and Amino Acids*, 2002.

BIBLIOGRAPHY

1. Eliasson, A. C. (Ed.) *Carbohydrates in Food.* Marcel Dekker Inc., 1996.

2. Szepesi, B. Carbohydrates. In E. E. Ziegler & L.J. Filer (Eds.) *Present Knowledge in Nutrition*, 7th Edition. Washington DC: ILSI Press, 1996.

FATS AND FATTY ACIDS

Within western culture, fats are typically associated negatively with chronic conditions such as high blood pressure, diabetes, cardiovascular disease, high cholesterol, and obesity. Even though too much fat does have negative effects on the body, fats are important to the body's normal functions.

Fat provides the body with energy. It is the major store of energy and serves to cushion and protect the internal organs. Unless there is fat in the diet, the body cannot absorb fat-soluble vitamins. In addition to the physiological role of fat, fats give food flavor and contribute to our enjoyment of food.

Fat makes up a substantial portion of the body's energy intake. It is recommended that the diet be no more than 30% fat[13].

No More Than
30% of Total Calorie Intake
Should Come From Fats

WHAT IS FAT?

Fat has various functions in the body. It is an alternative energy source to carbohydrates. Fat also provides the body with essential fatty acids, and dietary fat helps the body absorb fat-soluble vitamins. Fats, commonly referred to as **lipids**, may contain many different **fatty acids**. Essential fatty acids are used as structural components of cell membranes.

Fatty acids are classified according to their degree of **saturation**, a term referring to the number of **double bonds** a fat contains. Fats can be saturated, monounsaturated, or polyunsaturated:

- **Saturated fats** contain fatty acids with no double bonds and are generally straight-chained structures. Saturated fats contain mostly saturated fatty acids, while unsaturated fats contain mostly unsaturated fatty acids. Saturated fats are solid at room temperature—for example, butter.
- **Monounsaturated fats** contain fatty acids with a single double bond. Monounsaturated fats are liquid at room temperature (olive oil).
- **Polyunsaturated fats** contain fatty acids with multiple double bonds. Polyunsaturated fats are liquid at room temperature (many vegetable oils).

It is common to see fats referred to in the following manner, (16:0) for palmitic acid or (18:0) for stearic acid. The first number in parenthesis indicates the number of carbon atoms in the fatty acid and the second number indicates the number of double bonds.

ARE SOME FATS BETTER THAN OTHERS?

Monounsaturated and polyunsaturated fats are thought to lower **LDL cholesterol** levels. LDL cholesterol is the type of cholesterol that can promote coronary artery disease, increasing the risk of heart attack. Saturated fats contribute to coronary artery disease by elevating blood levels of LDL cholesterol which in turn promotes deposition of plaque in blood vessels in the heart, reducing blood flow and starving the heart for oxygen. The narrowed vessels are prone to complete blockage, causing a heart attack.

Diets worldwide contain a variety of types and levels of fat. However, none have received as much attention as the Mediterranean diet[3].

Although the Mediterranean diet is relatively high in fat—about 40%—it's low in saturated and polyunsaturated fats. The Mediterranean diet is high in fresh vegetables and fruits, and in olive oil.

Olive oil is high in Monounsaturated Fatty Acids (MFA). The health benefits of olive oil were originally attributed to the high content of monounsaturated fatty acids, specifically oleic acid.

However, there is conflicting evidence. Some researchers have found that monounsaturated fatty acids have a cholesterol lowering effect and that MFA increases **HDL**, the "good" cholesterol[5]. Others have found neutral results for **oleic acid**[7].

Yet others believe that the beneficial effects of olive oil are due to other factors in addition to oleic acid. Visioli and his research group[6] found that the **phenolic compound**s present in olive oil are powerful antioxidants. These researchers believe that the beneficial health effects of olive oil are attributable to a combination of oleic acid and phenolic compounds.

There has been much research effort and media attention focused on fatty acid composition in the food and in the diet. **Conjugated linoleic acid** is abundantly available in meat and dairy products. Conjugated linoleic acid, discovered in 1987, was found to be a potent inhibitor of mutagenesis—a process that may lead to cancer. Recent research has shown a possible role for conjugated linoleic acid in the prevention of cancer, diabetes and heart disease. However, further research is needed before we will fully understand the role conjugated linoleic acid plays in human health.

Omega-3 fatty acids have also received increasing attention. Omega-3 fatty acids are found in soybean oil, rapeseed oil, and more commonly in fish oil. Recently, eggs formulated to contain higher levels of omega-3 fatty acids[12] have become available to consumers. These have received much attention due to the triglyceride lowering effects of omega-3 fats[1].

Although research has demonstrated that many fatty acids have beneficial health effects, other fatty acids can cause or exacerbate health problems if they are consumed in excess.

Fatty acids are commonly hydrogenated during food processing to make the oils more shelf-stable. The **hydrogenation** process produces **trans fatty acids**, which are chemically similar to saturated fats. This process is commonly used to produce margarines and other vegetable fats that are solid or semi-solid at room temperature.

Because they behave much like saturated fatty acids, trans fatty acids have been shown to increase blood cholesterol and it is thought they may increase the risk of coronary heart disease. New classes of 'designer' margarines have been specifically formulated to lower blood cholesterol. These contain plant-derived **sterols** that inhibit the absorption of dietary cholesterol.

It was found that these sterol or stanol-containing oil products could lower LDL cholesterol within the body. In addition, sterols were able to decrease cholesterol in normal and high cholesterol subjects who ate a typical western diet[8,9,10,11].

Although few adverse effects have been noted for the ingestion of plant sterols and stanols, there have been indications of decreased absorption of several fat-soluble vitamins[10]. The American Heart Association recommends further studies to learn more about the effects of these designer oil products on normal and hypercholesterolaemic adults and children[11].

For additional information go to www.nap.edu and search for dietary reference intakes for energy, carbohydrates, fiber, fat, protein, and amino acids.

WHAT IS A FATTY ACID?

A fatty acid is made of a hydrocarbon chain with a carboxyl group at one end. Three major types of fatty acids are found in dietary fats: saturated, monounsaturated, and polyunsaturated.

A **saturated fatty acid** has a carbon chain that is completely and evenly filled, or saturated, with hydrogen atoms.

Carbon chains in **unsaturated fatty acids** are missing hydrogen atoms and have double bonds connecting the carbon atoms. The number of double bonds in the carbon chain determines what kind of unsaturated fatty acid it is.

Essential fatty acids play many roles in the body, including maintaining normal membrane function and serving as the starting material for important regulatory molecules such as prostaglandins.

FAT REQUIREMENTS

A normal healthy diet has no more than 30% of the total daily caloric intake from fat. No more than one third of total fat intake should be saturated fat.

Limit saturated fats to 1/3 of total fat intake!

Each gram of fat consumed is equal to 9 calories of energy, thus in a 2,400 calorie diet, a maximum of 720 calories or 80 grams should be from fat.

---

1 g fat = 9 calories

SOURCES OF FAT

Different types of fats come from different sources. This makes it relatively easy to select the types of fat we want to consume.

- **Saturated fats** are generally found in animal products of animal origin, such as meat and dairy products.
- **Monounsaturated fats** come from plant products such as peanut oil, olive oil, and canola oil.
- **Polyunsaturated fats** also come from plant products, including corn, soybeans, cottonseeds , safflowers, and sunflowers.

The most common fatty acids—palmitic acid, stearic acid, and oleic acid—are found in animal fats. The body can produce these fatty acids from sugar if we have a caloric surplus.

Essential fatty acids, on the other hand, are not produced by the body—we must get them from our food. Essential fatty acids are classified as either **omega-6** or **omega-3**, depending on the position of the carbon double bond. Omega-6 fatty acids are available from many different foods but omega-3 fatty acids are rarer.

Western diets typically are low in omega-3 fatty acids. The ratio of omega-6 to omega-3 fatty acids in a normal American diet is 3:1, but experts recommend a ratio of only 1:1[12]. Foods such as fish, certain vegetable oils, flax seeds, pumpkin seeds, walnuts, salmon, trout and tuna are all good sources of omega-3 fatty acids.

Try to eat similar amounts of omega-6 and omega-3.

Omega-6 fatty acids:

- Linoleic acid (LA)
- Arachidonic acid (AA)
- Gamma linolenic acid (GLA)
- Dihomogamma linolenic acid (DGLA)

Omega-3 fatty acids:

- Alpha linolenic acid (LNA)
- Eicosapentaenoic acid (EPA)
- Docosahexaenoic acid (DHA)

REFERENCES

1. Connor, W. E. (1986) Hypolipidemic effects of dietary omega 3 fatty acids in normal and hyperlipidemic humans: effectiveness and mechanisms. In A. P. Simopoulos, R. R. Kifer & R. E. Martin (Eds.), *Health effects of polyunsaturated fatty acids in seafoods.* Academic Press, Orlando, FL, pp. 173-210.

2. Seidelin, K. N., Myrup, B., and Fischer-Hansen, B. n-3 fatty acids in adipose tissue and coronary artery disease are inversely correlated. *American Journal of Clinical Nutrition* 55: 1117-9, 1992.

3. Keys, A. Mediterranean diet and public health: personal reflections. *Am J Clin Nutr* 1995; 61: 1321s -1323s.

4. Slatterly, M. L., Bensons, J., Ma, K. N., Schaffer, D. & Potter J. D. Trans fatty acids and colon cancer. *Nutr & Cancer* 2001; 39(2): 170-175.

5. Mensink, R. P., Katan, M. B. Effect of dietary fatty acids on serum lipids and lipoproteins. *Arterioscler Thromb Vasc Biol* 1992; 12:911-919.

6. Visioli, F., Poli, A., Galli, C. Antioxidant and other biological activities of phenols from olives and olive oil. *Medicinal Research Reviews* 2002; 22(1): 65-75.

7. Dougherty, R. M., Galli, C., Ferro-Luzzi, A. & Idcono, J. M. Lipid and phospholipid fatty acid composition of plasma red blood cells and platelets and how they are affected by dietary lipids: a study of normal subjects from Italy, Finland and USA. *Am J Clin Nutr* 1987; 45:443-455.

8. Miettinen, T. A., Puska, P., Gylling, H. et al. Reduction of serum cholesterol with sitostanol - ester margarine in a mildly hypercholesterolaemic population. *N Engl J Med* 1995; 333: 1308-1312.

9. Hallikainen, M. A., Sarkkinen, E. S., Erkkila, A.T. et al. Comparison of the effects of plant sterol ester and plant stanol ester enriched margarines in lowering serum cholesterol concentration in hypercholesterolaemic subjects on low-fat diet. *Eur J Clin Nutr* 2000; 54:715-725.

10. Hendriks, H. F., Westrate, J. A., Van Vliet, T. et al. Spreads enriched with 3 different levels of vegetable oil sterols and the degree of cholesterol lowering in normocholesterolaemic and mildly hypercholesterolaemic subjects. *Eur J Clin Nutr* 1999; 53:319-327.

11. Lichtenstein,A. H., Deckelbaum, R. J. Stanol/sterol ester containing foods and blood cholesterol levels. *Circulation* 2001; 103: 1177-1179.

12. Sims, J. S. Designer Eggs and their nutritional and functional significance. *World Rev. Nutr. Diet.* 1998;83:89-101.

13. Institute of Medicine. Food and Nutrition Board. *Dietary Reference intakes for Energy, Carbohydrates, Fiber, Fat, Protein and Amino Aicds,* 2002.

FIBER

Fiber is generally associated with bowel regularity and overall bowel health. However, fiber has many other positive health benefits as well. The importance of a high fiber diet has been continually stressed, both by the media and in scientific research.

WHAT IS FIBER?

Fruits, vegetables, and other plant foods contain structural components that are commonly referred to as fiber, roughage, or bran. It is commonly understood that the body does not digest fiber, but in fact fiber is a complex mixture of substances, some of which are slowly digested and absorbed.

There are two types of fiber: soluble and insoluble. **Soluble fiber** dissolves in water while **insoluble fiber** does not. Fiber is made up of a variety of components including cellulose, hemi-cellulose, lignin, pectin, mucilage, and gum[1].

WHAT DOES FIBER DO?

Fiber has many functions in promoting health. Different types of fiber are utilized by different parts of the body in different ways. Many of fiber's health-promoting effects are attributed to its ability to bind with other substances in the diet or secreted into the intestine via the bile. If the fiber binds with cholesterol or bile acids, blood cholesterol levels may be reduced.

In the large intestine, fiber can negatively or positively bind with chemicals that move through it. The large intestine comes in contact every day with mutagenic and carcinogenic compounds produced through digestion, ingestion or biotransformation by intestinal bacteria. Fiber can bind these carcinogenic and mutagenic compounds and prevent the contact of the carcinogen with the intestinal cells.

Recent studies have questioned the effectiveness of fiber supplements in reducing colon cancer risk[4,5,6]. It may be that the beneficial effects of fiber are realized only when it is eaten in whole foods as part of an overall healthy diet.

Recently, the United States Department of Agriculture (USDA) approved health claims that including oats as a part of a healthy diet reduces the risk of cardiovascular disease[3]. In particular, much attention has been focused on the soluble fibers **psyllium** and **ß glucan**.

One study demonstrated that the incorporation of soluble fiber into the diet resulted in a small but significant decrease in total cholesterol and in the ratio of LDL:HDL cholesterol [7]. According to the FDA, a person who eats at least 4 servings a day of foods containing the viscous fibers psyllium and oat ß glucan, in conjunction with a low-fat and cholesterol diet, can expect their serum lipids—and the risk of cardiovascular disease—to decrease. In effect, manufacturers of a food product may claim that it "will reduce the risk of coronary heart disease" if it contains at least 0.75g ß glucan or 1.78g of psyllium/serving[3]. Similar claims are currently undergoing review in Canada.

In addition, fiber adds bulk and softens the stool, which keeps the bowels working well, prevents constipation, and has been linked to a lowered risk of developing irritable bowel syndrome or colon cancer. Other reported benefits of fiber in the body are its role in the prevention of diverticular disease, Crohn's disease, and the formation of gallstones.

FIBER REQUIREMENTS

Unfortunately, the majority of North Americans do not eat the recommended levels of fiber. In the past, experts suggested that adults eat 30 to 35 grams of fiber a day[2]. However, more recent Adequate Intake

(AI) levels are 35 grams each day for men and 28 for women[8]. These levels are based on the amount of fiber that seems to protect against coronary heart disease.

Grams of Fiber Needed Each Day:

Men: 35g
Women: 28 g
Children 3–18: (Age + 5) grams

As a general rule of thumb, the amount of fiber needed is lower for children between the ages of 3 and 18. The amount of fiber a child needs each day is calculated by adding 5 to the child's age (e.g., an 11-year-old needs 16 grams of fiber per day).

Few people want to count grams of fiber. An easy way to decide how much fiber is enough is to eat the amount that maintains normal, regular bowel movements. A diet that includes 2 to 3 daily servings of whole grains, and at least 5 daily servings of fruits and vegetables, should provide a sufficient amount of fiber.

FIBER TOXICITY

There have been very few reports of ill effects related to fiber. It is possible that a large increase in fiber could reduce the absorption of essential minerals such as zinc, calcium, and iron. However, this is not likely to be a concern at the recommended levels of fiber intake.

It is important, though, to increase the amount of fiber in the diet gradually. Otherwise gas, abdominal bloating, and/or diarrhea could result.

Make sure to drink enough water as well, as water intake can influence the effects of increased fiber intake.

SOURCES OF FIBER

Fiber is found in foods of plant origin. Animal products do not contain fiber. Although most fruits, vegetables and grains contain insoluble fiber, soluble fiber is found in fruits such as apples and citrus, and in leafy green vegetables. Fruits and vegetables with edible skins are higher in fiber than those without. Oats, cereals, breads, and bran muffins are also good sources of fiber.

A nutritionally complete diet should contain all the nutrients and the amount of fiber necessary to maintain a healthy body. However, if your diet is deficient in fiber, supplements are available.

Food preparation techniques can also affect fiber content. For example, serving fruits and vegetables with edible skins and baking with whole grain flours will help you regularly include fiber in your diet.

Sources of Fiber[1]

Food	Serving size	Amount (grams)
Broccoli	½ c.	2.2
Brussels sprouts	½ c.	2.3
Carrots	½ c.	2
Celery	½ c.	1
Corn	½ c.	4
Corn on the cob	1 ear	5.9
Lettuce	1 c.	1
Canned peas	½ c.	4
Dried peas	½ c.	7.9
Spinach	1 c.	4
Lima, kidney, or baked beans	½ c.	10
Lentils	1 c.	8
Apple with peel	1 medium	3.5
Banana	1 medium	2.4
Grapefruit	½ medium	0.6
Orange	1 medium	2
Peach	1 medium	2
Strawberries	1 c.	3
Kiwi	1 medium	5
Pear	1 medium	4.5
Fiber One cereal	1 c.	14

Food	Serving size	Amount (grams)
100% Bran cereal	1 c.	13.5
All-Bran Extra Fiber cereal	1 c.	13
Raisin Bran cereal	1 c.	3.5
Whole wheat bread	1 slice	1.3
White, rye, or French bread	1 slice	0.7
Air-popped popcorn	3 ½ cups	4.5
Sunflower seeds	1 ounce	4

Source: Provisional table on the dietary fiber content of selected foods[1]

For additional information go to www.nap.edu and search for dietary reference intakes for energy, carbohydrates, fiber, fat, protein, and amino acids.

REFERENCES

1. Washington, D.C.: US Department of Agriculture, 1988. Available from:
http://www.indiadiets.com/foods/food_nutrients/Fibre.htm

2. Anon: American Gasteroenterological association medical position statement: Impact of dietary fiber on colon cancer occurrence. *Gastroenterology* 2001, 118: 1233-1234.

3. United States Food and Drug Administration. FDA final rule for federal labeling: health claims oats and coronary heart disease. *Federal Register* 1997; 62:3584-3681.

4. Schatzkin, A., Lanza, E., Corle, D., Lance, P., Iber, F., et al., Lack of effect of a low fat, higher fiber diet on the recurrence of colorectal adenomas. *New England Journal of Medicine*, 2000, 342: 1149-1155.

5. Alberts, D. S., Martinez, M. E., Rose, D. J., Guillen-Rodrigues, J. M., Marshall, J. R. et al., Lack of effect of a high fiber cereal supplement on the recurrence of a colorectal adenomas. *New Engl J Med*. 200; 342: 1156-1162.

6. Mason, J. G. Diet, Folate and Colon Cancer. *Curr Opin in Gastroenterol* 2002, 18: 229-234.

7. Jenkins, D. J. A., Kendall, C. W. C., Vuksan, V., Vidgen, E., Parker, T. et al., Soluble fiber intake at a dose approved by the US Food and Drug Administration for a claim of health benefits: serum lipid risk factors for cardiovascular diseases assessed in a randomized controlled cross over trial. *Am J Clin Nutr* 2002: 75: 834-839.

8. Institute of Medicine. Food and Nutrition Board. *Dietary Reference Intakes for Energy, Carbohydrates, Fiber, Fat, Protein and Amino Aicds,* 2002.

BIBLIOGRAPHY

1. http://www.indiadiets.com/foods/food_nutrients/Fibre.htm

2. Ziegler, E. & Filer, L. J. (Eds.) *Present Knowledge in Nutrition*, 7th edition. Washington DC: ILSI Press, 1996.

PROTEIN AND AMINO ACIDS

WHAT IS PROTEIN?

Protein is the dietary constituent that supplies amino acids, which the body assembles to make both structural and functional proteins essential to life. Although many amino acids are found in nature, only about 20 of these are used by the human body.

Various combinations of these amino acids make different proteins. The body can synthesize about half of the amino acids it needs. Most of the amino acids can be synthesized by the body, however ten cannot. The ten amino acids that the body can't synthesize are called the **indispensable amino acids**. These must be obtained through the diet.

WHAT IS THE FUNCTION OF PROTEIN?

Proteins serve two critical functions in the body:

- First, proteins are essential structural components of all tissues. For example, **collagen**, found in skin, is the most abundant protein in the body and serves a critical structural role. Hair and nails are primarily composed of protein. And muscle is made of specialized proteins called **contractile proteins** that work together to allow muscles to contract and relax.
- In addition to this structural role, other specialized proteins serve as **enzymes**—molecules that catalyze biological reactions in the

body. Enzymes are the engines of metabolism. They are the chemical machines that do everything, from taking food apart for energy to building new DNA for cell growth and reproduction.

Protein Requirements[1]

Age and gender	Daily protein requirement (g)
Children 1–3 years	23
Children 4–6 years	30
Children 7–10 years	34
Males 11–14 years	45
Males 15–18 years	56
Males 19-22 years	56
Males 23 years and older	56
Females 11-14 years	46
Females 15-18 years	46
Females 19-22 years	44
Females 23 years and older	44
Pregnant women	+30
Breastfeeding women	+20

HOW MUCH PROTEIN DO WE NEED?

Under normal circumstances, the body only requires enough protein to replace what is lost via normal metabolism. More protein is needed during times when there is a lot of growth, such as during childhood, pregnancy, exercising, and recovery from illness or injury. A healthy diet should have 10% to 35% of total calories from protein[4]. Absolute minimal protein needs are generally met by an intake of 0.5 gram protein (high quality) per kilogram of body weight. Typical western diets provide at least twice this amount of protein daily.

Each gram of protein contributes 4 calories of energy. Thus in a 2,400 calorie diet, 240 to 360 calories or 60 to 90 grams should be of high quality protein. See text box for sample calculations.

For additional information go to www.nap.edu and search for dietary reference intakes for energy, carbohydrates, fiber, fat, protein, and amino acids.

WHAT ARE THE EFFECTS OF PROTEIN DEFICIENCY?

Protein deficiency can cause abnormal growth and tissue development, particularly in the skin and nails. Muscle tone is also decreased in protein deficiency, and the body takes longer to recover from injuries and illnesses.

Children with protein deficiency will most likely not develop physically to their full potential. Severe protein deficiency can cause a disease called **Kwashiorkor**, which can be fatal. This disease causes stunted growth and mental development, swollen joints, and loss of hair color. This disease is more common in developing countries than in North America.

A rare disease called **phenlyketonuria, or PKU**, is caused by the body's inability to break down the amino acid **phenylalanine**. Phenylalanine builds up in the bloodstream, which in turn leads to buildup in the central nervous system, causing severe mental retardation. This disease is readily treatable through careful dietary management which limits phenylalanine intake.

Recent research into dairy foods offers hope for people who have PKU. It is possible to isolate a protein, commonly referred to as **glycomacropeptide**, which does not contain any phenylalanine[2]. Glycomacropeptide can be added to food products to serve as a protein source for individuals with PKU. Further research is under way[3].

SOURCES OF PROTEINS

Proteins can be obtained from foods of animal or plant origin. However, animal products generally contain more protein than plant products do.

Although proteins can be obtained from a variety of food products, not all proteins have the same nutritional value:

- **Complete proteins** contain a full complement of the indispensable amino acids—the amino acids the body can't manufacture. Eggs are considered to be a complete protein, and are generally used as a reference protein for comparing other protein sources.

- **Incomplete proteins** are generally missing one or more indispensable amino acids. The amino acid in lowest concentration is referred to as the **limiting amino acid**.
- **Complementary proteins** are two proteins with differing limiting amino acids that are consumed together to obtain the full complement of indispensable amino acids. For example, eating both corn and wheat protein would provide an overall dietary protein quality greater than that derived from either protein alone.

Incomplete and complete proteins are also commonly referred to as **low and high quality proteins**, respectively.

Animal products generally contain high quality proteins, while plants other than soy generally contain lower quality proteins. Soy protein is the only high quality protein found in plants. It is also possible to obtain sufficient high quality protein from plants by combining plant products with differing amino acid profiles.

Food sources of protein

Food	Amount of protein (g)
Chicken, 3 oz.	20
Ground Beef	21
Pork chop, 2 oz.	15
Milk, 1 c.	9
Egg, 1 large	6
Cheddar cheese, 1 oz.	7
Beans, ¾ c.	11
Peanut butter, 2 tbsp.	8
Bread, 1 slice	2
Rice, ½ c.	5
Nuts, 2 tbsp.	5
Soybeans, ½ c.	10
Ground beef, 3 oz.	21

REFERENCES

1. *Recommended Dietary Allowances*, Ninth Edition. National Academy of Sciences. Washington, D.C. 1980.

2. Nakeano, T., Silva-Hernandez, E. R., Ikawa, N. & Ozmek, L. Purification of Kappa-casein glycomacropeptide from sweet whey with undetectable levels of phenylalanine. *Biotechnol prog.* 2002; 18(2):409-412.

3. Ozimek, Lech, personal communication.

4. Institute of Medicine. Food and Nutrition Board. *Dietary Reference intakes for Energy, Carbohydrates, Fiber, Fat, Protein and Amino Aicds,* 2002.

VITAMINS

BIOTIN (VITAMIN H)

WHAT IS BIOTIN?

Biotin, also known as **Vitamin H**, is a water-soluble vitamin[8] discovered in 1936. Biotin is an essential nutrient that can be found in almost all plant and animal cells.

WHAT DOES BIOTIN DO?

Biotin is an essential cofactor for special enzymes, called **carboxylases**, that speed up essential metabolic processes. These processes are necessary to synthesize fatty acids, form glucose, amino acids and fats, metabolize **leucine** (an essential amino acid), and metabolize amino acids, cholesterol, and odd fatty acids. Biotin also helps the body use B vitamins[3].

Biotin intake levels by age

Age[3]	Adequate intake (µg/day)
Children 0–6 months	5
Children 7–12 months	6
Children 1–3 years	8
Children 4–8 years	12
Children 9–13 years	20
Teenagers 14–18 years	25
Adults 19 years and older	30
Pregnant women of any age	30
Breastfeeding women of any age	35

An increased need for biotin is often associated with pregnancy, long-term use of anti-seizure drugs, and some inherited disorders.

WHAT CAUSES A BIOTIN DEFICIENCY?

Biotin deficiency is not common. It can result from intravenous feeding with inadequate biotin or when biotin intake is not increased when the body's need for biotin rises (as for example in pregnancy).

In certain people, such as weight lifters, excessive consumption of raw egg white can impair biotin absorption. Raw egg whites contain a **glycoprotein** called **avidin** which can bind irreversibly with biotin, decreasing the amount of biotin available to the body. This binding can be avoided easily, by simply cooking the eggs[2,3].

BIOTIN TOXICITY

There have been no reports of biotin toxicity.

SOURCES OF BIOTIN

It is important to include a dietary source of biotin in your diet. Every living organism needs biotin, but the only organisms capable of synthesizing it are bacteria, yeasts, molds, algae, and some plants[2]. Bacteria in the large intestine produce negligible amounts of biotin.

Many foods contain biotin. However, the amounts are usually small, and cooking or preserving food decreases the amount of biotin available for absorption. Egg yolk, liver, yeast, and wheat bran are some of the best food sources of biotin.

Food sources of biotin[1]

Food	Amount of biotin (μg)
Bakers yeast, 7 g	14
Wheat bran, 1 oz.	14
Whole wheat bread, 1 slice	6
Cooked egg, 1 large	25
Camembert cheese, 1 oz.	6
Cheddar cheese, 1 oz.	2
Cooked liver, 3 oz.	27
Cooked chicken, 3 oz.	3
Cooked pork, 3 oz.	2
Cooked salmon, 3 oz.	4
Avocado, 1 whole	6
Raspberries, 1 c.	2

REFERENCES

1. Food and Nutrition Board, Institute of Medicine. Biotin. *Dietary Reference Intakes: Thiamin, Riboflavin, Niacin, Vitamin B-6, Vitamin B-12, Pantothenic Acid, Biotin, and Choline*. Washington, D.C.: National Academy Press; 1998:374-389.

2. Mock, D. M. Biotin. In E. E. Ziegler & L. J. Filer (Eds.) *Present Knowledge in Nutrition*. 7th ed. Washington D.C.: ILSI Press; 1996:220-236.

3. Linus Pauling Institute. Oregon State University. http://lpi.orst.edu/

CHOLINE

WHAT IS CHOLINE?

Functionally, choline is similar to B Vitamins. Choline acts as a **coenzyme** during metabolism and is required for many other essential processes within the body.

WHAT DOES CHOLINE DO?

Choline plays an important role in the body's processing of fats and cholesterol.

Choline and the compounds derived from it are used in the creation of cell membrane structural components, neurotransmitters, and cell signaling molecules. The body uses choline to form **betaine**, a substance that reduces the risk of heart disease by converting **homocysteine** to **methionine**.

Another important function of choline is the synthesis of the neurotransmitter **acetylcholine**. Acetylcholine is important to brain processes such as learning and memory.

It is recommended that women consume 425 mg of choline daily and men 550 mg[1].

Daily Choline Requirement:
Women: 425 mg
Men: 550 mg

WHAT ARE THE EFFECTS OF CHOLINE DEFICIENCY?

Choline deficiency may result in memory and learning disorders, as the body will not be able to produce sufficient acetylcholine. Drugs that block the synthesis or interaction of acetylcholine with receptor sites are known to have a negative effect on memory in normal individuals[2,3].

Choline deficiency may also cause a condition known as "fatty liver." Because choline is important to the processing of fat, a lack of choline causes fat to build up in and damage the liver. Choline deficiency may also increase the risk of hardened arteries and high blood pressure.

Sources of choline[4]

Food	Amount of choline (mg)
Egg, 1 large	200–300
Beef liver, 3 oz.	453
Beef, 3 oz.	59
Cauliflower, 1 c.	55
Peanut butter, 2 T.	26
Grape juice, 8 oz.	13
Baked potato, 1 medium	18
Whole milk, 8 oz.	10
Tomato, 1 medium	7
Orange, 1 medium	10
Whole wheat bread, 1 slice	4

REFERENCES

1. The Food and Nutrition Board. Institute of Medicine, Food and Nutrition Board. *Dietary Reference Intakes: Thiamin, Riboflavin, Niacin, Vitamin B-6, Vitamin B-12, Pantothenic Acid, Biotin, and Choline*. Washington, DC: National Academy Press, 1998: 390-422.

2. Holmes, G. Z., Yang, Y., Liu, Z., Cermak, J. M., Sarkisian, M. R., Stafstrom, C. E. et al. Seizure induced memory impairment is reduced by choline supplementation before or after status epilepticus. *Epilepsy Research* 2002; 48(1-2): 3-13.

3. Nakamura, A., Suzuki, Y., Umegaki, H., Ikari, H., Tajima, T., Endo, H. & Iguchi, A. A dietary restriction of choline reduces hippocampal acetylcholine release in rats: in vivo microdialysis study. *Brain Research Bulletin* 2001; 56(6): 593-597.

4. Linus Pauling Institute. Oregon State University. http://lpi.orst.edu/

BIBLIOGRAPHY

1. Zeisel, S.H. Choline: an essential nutrient for humans. *Nutrition*. 2000; volume 16: 669-671.

2. Linus Pauling Institute. Oregon State University. http://lpi.orst.edu/

FOLIC ACID (FOLATE)

WHAT IS FOLIC ACID?

Folic acid, also known as **folate**, is a water-soluble, essential B vitamin.

WHAT ARE THE BENEFITS OF FOLIC ACID?

Folic acid has many health benefits, including the prevention of neural tube defects such as **spina bifida** in infants[8,9,10]. Adequate dietary folate can reduce the risk of birth defects by up to 70%. Because folate is so important in preventing neural tube defects, the US government has made it mandatory for all cereal grains to be fortified with 140 µg of folic acid for every 100 g of cereal grain.

In addition to the prevention of neural tube defects, folic acid has been found to reduce the risk of certain cancers and heart disease. A study conducted between 1996 and 1998 by a team of American and Chinese researchers showed that folic acid can prevent breast cancer in pre- and post-menopausal women. Women taking more than 345 µg of folic acid per day reduced their risk of breast cancer by 38% compared to women taking less than 195 µg per day[1]. In several studies, high consumption of folate was associated with a decreased risk of breast cancer among women with regular alcohol consumption. There was not a similar protective mechanism among women who did not drink alcohol[5,6,7].

The use of folic acid supplements has been found to effectively reduce the rate and severity of **ischemic heart diseases** by lowering blood levels of homocysteine. People with ischemic heart disease can benefit from taking 800 μg of folic acid per day[2,4]. Moderate to high intakes of folate have also been shown to decrease the incidence of acute coronary events in men[3].

FOLIC ACID REQUIREMENTS

The requirement for folic acid is 400μg per day for the average adult. To prevent neural tube defects, women of childbearing age should consume an additional 400 μg of folic acid daily[11].

WHAT CAUSES FOLIC ACID DEFICIENCY?

The most common cause of folic acid deficiency is inadequate dietary intake. Inadequate consumption of fresh fruits and vegetables or consumption of only cooked vegetables can contribute to a deficiency.

Cooking destroys folic acid. Certain medicines may also contribute to folic acid deficiency by interfering with the metabolism of folic acid. Pregnancy, breastfeeding, and drinking alcoholic beverages are all conditions that increase the body's requirement for folic acid, thereby increasing the risk of inadequate dietary intake.

WHAT ARE THE EFFECTS OF FOLIC ACID DEFICIENCY?

Folic acid deficiency results in high blood levels of **homocysteine,** which increases the risk of heart disease. Other possible effects of deficiency include headaches, memory loss, loss of appetite, weight loss, weakness, irritability, paranoia, insomnia, and a sore red tongue. Folic acid deficiency may also cause anemia due to the role folic acid plays in red blood cell production.

In pregnant women, folic acid deficiency increases the likelihood of serious birth defects to the spinal cord and endangers the infant's heart development.

SOURCES OF FOLIC ACID

Natural folic acid is found in leafy green vegetables such as broccoli and spinach, as well as in some fruit and juices. Synthetic folic acid is found in breads and cereals, and in multivitamins.

Food Sources of Folic Acid

Vegetables	Amount of folic acid (μg)
Asparagus, 1 c.	176
Cooked black beans, ½ c.	128
Cooked white beans, ½ c.	123
Beets, 1 c.	90
Brussels sprouts, 1 c.	94
Cantaloupe, 1 c.	150
Cauliflower, 1 c.	63
Cooked lentils, 1 c.	179
Romaine lettuce, 1 c.	76
Peas, 1 c.	94
Cooked spinach, 1 c.	131
Raw spinach, 1 c.	108
Fruits	
Orange juice, 1 c.	109
Pineapple juice, 1 c.	58
Raspberries, 1 c.	49
Strawberries, 1 c.	65
Cereals	
Grape-nuts 1 c.	402
All-bran 1 c.	301
Bran flakes 1 c.	173
Wheat germ 1/3 c.	108
Wheaties, 1 c.	102

Nuts	
Peanuts, 1/3 c.	117
Sunflower seeds, 1/3 c.	109
Meats	
Cooked chicken liver, ¼ c.	269
Cooked beef liver, 4 oz.	162

REFERENCES

1. Martha, J. et al., Dietary folate intake and breast cancer risk; results from the Shanghai breast cancer study. *Cancer Research* 2001; 61: 7136-7141.

2. Wald, D. S. et al., Randomized trial of folic acid supplementation and serum homocysteine levels. *Archives of Internal Medicine* 2001; 161: 695-700.

3. Voutilainen, S., Rissanen, T. H., Virtanen, J., Lakka, T. A., Salonen, J. T. Low dietary folate intake is associated with an excess incidence of acute coronary events: Te Kuopio Ischemic Heart Disease risk factor study. *Circulation*. 2001: 103: 2674-2680.

4. El-Khairy, L., Ueland, P. M., Refsum, H., Graham, I. M. & Vollset, S. E. Plasma total cysteine as a risk factor for vascular disease: The European concerted action project. *Circulation*. 2001; 103: 2544-2549.

5. Zhang, S., Hunter, D. J., Hankinson, S. E. et al. A prospective study of folate intake and the risk of breast cancer. *JAMA*. 1999; 281:1632-1637.

6. Rohan. T. E., Jain, M. G., Howe, G. R. & Miller, A. B. Dietary folate consumption and breast cancer risk. *J Natl Cancer Inst*. 2000; 92: 266-269.

7. Sellers T. A., Kushi, L. H., Cerhan, J. R. et al. Dietary folate intake, alcohol and risk of breast cancer in prospective study of post menopausal women. *Epidemiology*. 2001; 12:420-428.

8. Milunsky, A., Jick, H., Jick, S. S. et al. Multivitamin/folic acid supplementation in early pregnancy reduces the prevalence of neural tube defects. *JAMA*. 1989; 262:2847-2852.

9. Czeizel, A. E. & Dudas, I. Prevention of the first occurrence of neural tube defects by periconceptional vitamin supplementation. *N Engl J Med*. 1992; 327: 1832-1835.

10. MRC Vitamin Study Research Group. Prevention of neural tube defects: results of the Medical Research Council vitamin Study. *Lancet*. 1991; 338:131-137.

11. Institute of Medicine. Food and Nutrition Board. *Dietary Reference Intakes for Thiamin, Riboflavin, Niacin, Vitamin B6, Folate, Vitamin B12, Pantothenic Acid, Biotin, and Choline*. 1998.

PANTOTHENIC ACID (VITAMIN B5)

The name pantothenic comes from the Greek word pántothen, which means "from all quarters." This name is appropriate, as pantothenic acid is used in many metabolic reactions throughout the body.

WHAT IS PANTOTHENIC ACID?

Pantothenic acid, also known as **Vitamin B5**, is an essential B-complex vitamin used in growth, reproduction, and normal body functions. Pantothenic acid is also found in two alternate forms, **panthenol** and **calcium pantothenate**.

This B vitamin is of particular importance to the body because it is an integral component of two important coenzymes: **coenzyme A** and **acyl**

carrier protein. These two enzymes are involved in a diverse array of metabolic reactions within the body[6].

WHAT DOES PANTOTHENIC ACID DO?

Pantothenic acid has many uses in the body. It is required for the metabolism and liberation of energy from proteins, fats, and carbohydrates. It also plays an important role in the synthesis of many **cofactors**, tissue components, enzymes, antibodies, hormones and neurotransmitters. Pantothenic acid plays an integral role in the functioning of the nervous system, adrenal gland, immune system, and general body maintenance[6].

Pantothenic acid requirements[7]

Age	Recommended daily intake (mg)
Children 0–6 months	2
Children 6 months–3 years	3
Children 4–6 years	3–4
Children 7–10 years	4–5
Children 11 years and older and adults	4–7
Women taking birth control pills	May require more than 4–7.

WHAT ARE THE EFFECTS OF PANTOTHENIC ACID DEFICIENCY?

Pantothenic acid deficiency does not commonly occur in people. However, in experimental animals given a diet containing no pantothenic acid, there have been observations of impaired growth along with reduced food intake, and systemic functional impairments often resulting in death.

In humans, pantothenic acid deficiency may cause depression, changes in personality, fatigue, stomach aches and vomiting, insomnia, and a burning sensation in the feet[1,2,3].

PANTOTHENIC ACID TOXICITY

Diarrhea has been caused by doses of 10 to 20 grams of pantothenic acid a day.

SOURCES OF PANTOTHENIC ACID

Pantothenic acid is found in many foods, such as organ meats, poultry, lobster, soybeans, lentils, split peas, yogurt, avocado, mushrooms, sweet potatoes, brewer's yeast, and wheat germ. Pantothenic acid supplements are rarely needed, since a varied diet will provide enough pantothenic acid for most people.

Food Sources of pantothenic acid[6]

Food	Amount of pantothenic acid (mg)
Cod fish, 3 oz.	0.15
Tuna, 3 oz.	0.18
Cooked chicken, 3 oz.	0.98
Cooked egg, 1 large	0.61
Milk, 1 c.	0.79
Yogurt, 8 oz.	1.35
Steamed broccoli, ½ c.	0.40
Cooked lentils, ½ c.	0.64
Split peas, ½ c.	0.59
Avocado, 1 whole	1.68
Sweet potato, ½ c.	0.74
Raw mushrooms, ½ c.	0.51
Cooked lobster, 3 oz.	0.24

REFERENCES

1. Song, W. O. Pantothenic acid: How much do we know about this b vitamin? *Nutr Today*. 1990; 25: 19-26.

2. Annous, K. F.& Song, W. O. Pantothenic acid uptake and metabolism by the red blood cell. *J Nutr*. 1995; 125: 2586-2593.

3. Tahiliani, A. G. & Beinlich, C. J. Pantothenic acid in health and disease. *Vitamins and Hormones*. 1990; 46: 165-228.

4. Zempleni, J., Stanley, J. & Mock, D. M. Proliferation of peripheral blood mononuclear cells causes increased expression of the sodium-dependent multivitamin transporter gene and increased uptake of pantothenic acid. *Journal of Nutritional Biochemistry*, 2001; 12(8): 465-473.

5. Jakus, J., Kriska, T. & Rozalia, V. Effect of multivitamins in an effervescent preparation on the respiratory burst of peritoneal macrophages in mice. *British Journal of Nutrition*, 2002; 87: 501-508.

6. Linus Pauling Institute. Oregon State University. http://lpi.orst.edu/

7. Institute of Medicine. Food and Nutrition Board. *Dietary Reference Intakes for Thiamin, Riboflavin, Niacin, Vitamin B6, Folate, Vitamin B12, Pantothenic Acid, Biotin, and Choline*. 1998.

BIBLIOGRAPHY

1. Plesofsky-Vig, N. Pantothenic acid. In Shils, M. et al. (Eds.) *Nutrition in Health and Disease*, 9th Edition. Baltimore: Williams & Wilkins, 1999: pp. 423-432.

2. Brody, T. *Nutritional Biochemistry*. San Diego, CA: Academic Press, 1999: pp. 613-617.

3. Plesofsky-Vig, N. Pantothenic acid. In E. E. Ziegler & L.J. Filer (Eds.) *Present Knowledge in Nutrition*, 7th Edition. Washington DC: ILSI Press, 1996: pp. 237-244.

VITAMIN A

Vitamin A is an essential vitamin required for growth, reproduction, and visual function. Vitamin A activity is derived from both pre-formed vitamin A (**retinol**) and **pro-vitamin A carotenoids**. Cells lining the intestine can convert pro-vitamin A carotenoids into retinol, the form of Vitamin A that the body uses[2].

Carotenoids are commonly associated with yellow, orange, or red foods. Among the pro-vitamin A carotenoids, **beta-carotene** provides the most vitamin A activity and is the most abundant in a typical diet.

WHAT DOES VITAMIN A DO?

Vitamin A is probably best known for its essential role in vision and the prevention of night blindness. Vitamin A is also used in the growth, repair, and maintenance of many body parts and tissues. Vitamin A keeps the skin soft and smooth, helps form bones and teeth, and protects the mucous membranes in the respiratory tract from infections and air pollutants.

Recent medical studies have linked diets high in beta-carotene to reduced risk of lung and oral cancers. It is believed that Vitamin A plays a role in cancer prevention due to its role in enhancing positive cell growth and its role in the maintenance of **epithelial cell integrity**.

Vitamin A Requirements[1]:

Age (years)	Gender	RDA (µg/day)
1–3	Males and females	300
4–8	Males and females	400
9–13	Males and females	600
14–18	Males	900
14–18	Females	700
14–50	Pregnant females	750–770
14–50	Lactating females	1200–1300
19+	Males	900
19+	Females	700

WHAT ARE THE EFFECTS OF VITAMIN A DEFICIENCY?

One of the earliest symptoms of Vitamin A deficiency is night blindness. Vitamin A deficiency is quite common in developing countries; it is estimated that 50,000 people go blind every year as a result.

Other effects of Vitamin A deficiency include dry, rough, and/or scaly skin, a reduced sense of smell and appetite, weight loss, tiredness, stunted growth, and problems with gums and teeth.

Researchers have found that children with Vitamin A deficiency have a higher incidence of gastrointestinal related disorders such as diarrhea as well as respiratory infections[2]. As a result, Vitamin A deficiency contributes to a high infant mortality rate. Researchers have found that adding Vitamin A supplements to the diet can significantly reduce childhood mortality[5,6].

VITAMIN A TOXICITY

Dietary vitamin A overload, referred to as **hypervitaminosis A,** is rare but it can occur from excess consumption of foods rich in pre-formed vitamin A, such as animal liver. Symptoms of hypervitaminosis A are nausea, dizziness, blurred vision, headaches, and a lack of muscular coordination[1,3,4].

The major effects of hypervitaminosis A are birth defects, liver abnormalities, and reduced bone mineral density. Reduced bone mineral density could result in **osteoporosis**[1].

REQUIREMENTS FOR AND SOURCE S OF VITAMIN A

For someone consuming 2000 kcal/day the recommended intake of vitamin A is 5000 IU.

Vitamin A precursors (that is, compounds that can be transformed to Vitamin A within the body) such as **retinyl esters** and **carotenoids** are commonly found in foods. Retinyl esters are found in egg yolks, butter, whole milk products, liver, and fish. Carotenoids are abundantly available in brightly colored yellow or orange vegetables and fruits, as well as in green vegetables.

Sources of Vitamin A

Food	International units (IU)	% DV
Beef liver, cooked, 3 oz.	30,325	610
Chicken liver, cooked, 3 oz.	13,920	280
Egg substitute, fortified, ¼ c.	1,355	25
Fat free milk, fortified, 1 c.	500	10
Whole milk, 1 c.	305	6
Cheddar cheese, 1 oz.	300	6
Swiss cheese, 1 oz.	240	4
Whole egg, 1 medium	280	6
Low-fat yogurt, 1 c.	120	2
Carrot, 1 raw, 7.5 in.	20,250	410
Carrots, boiled, ½ c.	19,150	380
Carrot juice, canned, ½ c.	12,915	260
Mango, 1	8,050	160
Sweet potatoes, ½ c.	7,430	150
Boiled spinach, ½ c.	7,370	150
Raw spinach, 1 c.	2,015	40
Cantaloupe, 1 c.	5,160	100
Broccoli, ½ c.	1,740	35
Tomato juice, 6 oz.	1,010	20
Raw peach, 1 medium	525	10
Raw papaya, 1 small	430	10
Raw orange, 1 large	375	8

REFERENCES

1. Institute of Medicine, Food and Nutrition Board. *Dietary Reference Intakes: Vitamin A, Vitamin K, Arsenic, Boron, Chromium, Copper, Iodine, Iron, Manganese, Molybdenum, Nickel, Silicon, Vanadium, and Zinc*. National Academy Press, Washington, DC, 2001.

2. Linus Pauling Institute. Oregon State University. http://lpi.orst.edu/

3. Bendich, A & Langseth, L. Safety of Vitamin A. *Am J Clin Nutr* 1989;49:358-371.

4. Udall, J. N., Greene, H. L. Vitamin update. *Pediatrics in Review* 1992;13:185-194.

5. West, K., Pokhrei, R., Katz, J. et al., Efficacy of Vitamin A in reducing preschool child mortality in Nepal. *Lancet* 1992; 338: 67-71.

6. Ghana VAST Study Team. Vitamin A supplementation in Northern Ghana: Effects on clinic attendances, hospital admissions and childhood mortality. *Lancet* 1993, 342: 7-12.

VITAMIN B1 (THIAMIN)

WHAT IS VITAMIN B1?

Vitamin B1 is also called **thiamin**. It is a water-soluble vitamin that plays an important role in catalyzing energy-releasing reactions within the body.

It is commonly found in the body as **thiamine pyrophosphate** or **TPP**. This component is a co-factor for several enzymes involved in energy metabolism[4]. The body needs Vitamin B1 to metabolize carbohydrates, proteins, and fats. Vitamin B1 is also a necessary component in the formation of **ATP**, and helps keep nerve cells and the nervous system functioning properly.

WHAT DOES VITAMIN B1 DEFICIENCY DO?

Deficiency of thiamin causes a condition called **beriberi**. Symptoms of beriberi include fatigue, irritability, memory loss, problems with sleep, chest pain, abdominal pain, constipation, and anorexia. Inadequate thiamin intake may also cause nervousness, weak and sore muscles, numb hands and feet, and loss of coordination.

Adding thiamin supplements to the diet has been found to halt mental deterioration. Recently, researchers have noted correlations between thiamin deficiency and brain and memory function. Wernicke-Korsakoff syndrome is a form of brain damage that results in amnesia and dementia. The onset of Wernicke-Korsakoff syndrome is brought about by long-term alcoholism or alcohol abuse.

CAN VITAMIN B1 CAUSE ANY BAD EFFECTS?

Since the body is able to excrete excess Vitamin B1 in the urine, there are rarely any toxic effects of overconsumption. Vitamin B1 toxicity usually occurs only where supplements over 3 grams a day are taken for a long period of time.

VITAMIN B1 REQUIREMENTS

The general recommendation for Vitamin B1 is 0.5 mg per every 1000 calories of carbohydrates eaten. The recommended daily allowance for Vitamin B1 for women increases during pregnancy and lactation. Because thiamin plays an integral role in converting carbohydrates to energy, adding carbohydrates to the diet raises the body's need for thiamin.

Many factors such as alcohol consumption, antacids, birth control pills, or hormone replacement therapy can lower the level of Vitamin B1 in the body. Alcoholism, malabsorption diseases, or nutritionally poor diets commonly are all associated with thiamin deficiency.

WHERE DO WE GET VITAMIN B1?

The body does not store Vitamin B1 in large amounts, so it is important to include adequate sources in your diet. Pork, red meats, organ meats, poultry, fish, whole grains, nuts, beans, peas, milk, cauliflower, asparagus, and spinach are all good sources of Vitamin B1.

REFERENCES

1. Butterworth, R. F., et al. Thiamine deficiency in AIDS. *Lancet* 1991;338:1086.

2. Gold, M., Hauser, R. A. & Chen, M. F. Plasma thiamin deficiency associated with Alzheimer's disease but not Parkinson's disease. *Metabolic Brain Disease* 1998, 13(1): 43-53.

3. Rodriguez-Martin, J. L., Qizilbash, N. & Lopez-Arrieta, J. M. Thiamin for Alzheimer's disease. *Cochrane Database of Systemic Review* 2001(2) CD001498.

4. Linus Pauling Institute. Oregon State University. http://lpi.orst.edu/

BIBLIOGRAPHY

1. Ziegler, E. & Filer, L. J. (Eds.) *Present Knowledge in Nutrition*, 7th edition. Washington DC: ILSI Press, 1996.

VITAMIN B2 (RIBOFLAVIN)

Vitamin B2 or **riboflavin** serves many functions in the body. It is required for the metabolism and release of energy from amino acids, fatty acids, fats, proteins, and carbohydrates. Vitamin B2 also plays an important role in the function of the immune system by supporting the formation of red blood cells and antibodies. Vitamin B2 also has antioxidant properties.

Vitamin B2 contributes to the health of our eyes by relieving watery eye fatigue and possibly helping to prevent and treat cataracts. The mucous membranes in the digestive tract need Vitamin B2 to stay healthy and to absorb enough iron and Vitamin B6. Our skin, hair, nails, and joints need Vitamin B2 for reproduction, repair, and growth.

Vitamin B2 is water-soluble and is not stored in body fat. Excess Vitamin B2 is excreted in the urine.

WHAT ARE THE EFFECTS OF A VITAMIN B2 DEFICIENCY?

Vitamin B2 deficiency may cause the mouth and tongue to become swollen, and cracks and sores to form in the corners of the mouth. Other possible effects are skin lesions, dermatitis, burning feet, maldigestion, insomnia, light sensitivity, hair loss, dizziness, stunted growth, and slow mental response[3].

VITAMIN B2 TOXICITY

Because Vitamin B2 is water soluble, any excess is excreted in the sweat and urine. This makes it very unlikely for a toxic level of Vitamin B2 to build up in the body. Toxicity due to excess B2 intake is rare, although high doses have been associated with numbness, itching, burning sensations, and/or sensitivity to light.

VITAMIN B2 REQUIREMENTS

Recommended requirements for Vitamin B2 vary considerably, and are listed in the table below. Vitamin B2 requirements increase during periods of rapid growth and during periods of high physical activity. Requirements are also higher for individuals whose diet is rich in protein. Additional Vitamin B2 may be needed if there is excessive alcohol consumption, or if antibiotics or birth control pills are taken.

RDA for Vitamin B2[1,2]

Age (in years) and gender	RDA (mg/day)
Children 0–1	0.4
Children 1–3	0.8
Children 4–6	1
Children 7–9	1.3
Boys 10–12	1.4
Girls 10–12	1.3
Boys 13–15	1.6
Girls 13–15	1.4
Boys 16–19	1.6
Girls 16–19	1.5
Men 19 and older	1.6
Women 19 and older	1.5
Pregnant women	1.6
Breastfeeding women	1.8

SOURCE S OF VITAMIN B2

Riboflavin is found in a variety of foods, including most plant and animal products. However, organ meats, nuts, cheese, brewer's yeast, eggs, milk, and lean meat are the best sources of Vitamin B2. Other good sources include mushrooms, soybeans, whole grains, leafy green vegetables, spinach, legumes, fish, and yogurt.

REFERENCES

1. Food and Nutrition Board, Institute of Medicine. *Dietary Reference Intakes for Thiamin, Riboflavin, Niacin, Vitamin B6, Folate, Vitamin B12, Pantothenic Acid, Biotin, and Choline*. Washington, DC: National Academy Press; 1998.

2. Recommended Dietary Allowances, 10th ed. Washington, D.C.: National Academy Press, 1989.

3. Linus Pauling Institute. Oregon State University. http://LPI.orst.edu/

BIBLIOGRAPHY

1. Ziegler, E. & Filer, L. J. (Eds.) *Present Knowledge in Nutrition*, 7th edition. Washington DC: ILSI Press, 1996.

VITAMIN B3 (NIACINAMIDE OR NIACIN)

WHAT IS VITAMIN B3?

There are two forms of this water-soluble vitamin: **niacin** and **niacinamide**. Some important differences between the two forms are discussed below.

WHAT DOES VITAMIN B3 DO?

Vitamin B3 helps keep the skin healthy, the blood circulating properly, the cholesterol levels low (only the niacin form of Vitamin B3 affects cholesterol levels), and helps in nervous system functioning. Vitamin B3 is also used in the metabolism of carbohydrates, fats, and proteins. It is necessary for releasing energy, for transforming carbohydrates to fats, and for the processing of alcohol. Vitamin B3 also plays a part in the secretion of bile and stomach juices, and in the synthesis of sex hormones.

VITAMIN B3 TOXICITY

There are no documented cases of niacin toxicity from food. However, pharmacologic doses of niacin can lead to adverse effects[3]. It is generally safe to take supplements of the niacinamide form of Vitamin B3 as long as the amount taken is less than 1,000 mg per day. Rare liver problems have been reported where doses exceeded 1,000 mg per day.

It is important that you consult a medical doctor before you begin taking high doses of niacin. As little as 50 to 100 mg of the niacin form can cause flushing, headache, and stomach aches. Although doses of up to

3000 mg per day of niacin are sometimes prescribed by doctors to lower cholesterol levels, doses this high can cause damage to the liver and the eyes, can cause diabetes and gastritis, and can elevate the level of uric acid in the blood.

VITAMIN B3 REQUIREMENTS[4]

The requirements for B3 vary with age and gender and are listed in the table below.

RDA for Vitamin B3

Age and gender	RDA (mg/day)
Children 0–6 months	2
Children 6 months–1 year	4
Children 1–3 years	10
Children 4–8 years	15
Males 9–13 years	12
Males 14 years and older	16
Females 9–13 years	12
Females 14 years and older	14
Pregnant women	18
Breastfeeding women	17

ARE THERE SAFER ALTERNATIVES?

Many dietary supplements are currently available on the market. **Time-release** niacin supplements have fewer side effects, but can nevertheless have damaging effects on the liver[1]. New **extended-release** or **partial-release** forms of niacin have been found to be tolerated well and, most importantly, have fewer side effects than traditional niacin supplements[2].

SOURCES OF VITAMIN B3

High-protein foods such as beef, brewer's yeast, chicken, beans, peas, fish, liver, peanuts and peanut butter, pork, potatoes, low-fat dairy products, soybeans, wheat germ, and nuts are all excellent sources of Vitamin B3. Fortified cereals can provide between 20 and 27 mg of niacin per one-cup serving.

Oranges are the only fruits that contain Vitamin B3. In addition, Vitamin B3 is found in herbs such as alfalfa, burdock root, cayenne, chamomile, chickweed, eyebright, fennel seed, hops, licorice, parsley, peppermint, and rose hips.

REFERENCES

1. McKenney, J. M., Proctor, J. D., Harris, S., Chinchili, V. M. A comparison of the efficacy and toxic effects of sustained vs. immediate-release niacin in hypercholesterolemic patients. *JAMA* 1994;271:672-7.

2. Goldberg, A., Alagona, P. Jr., Capuzzi, D. M., et al. Multiple-dose efficacy and safety of an extended-release form of niacin in the management of hyperlipidemia. *Am J Cardio*l 2000;85:1100-5.

3. Linus Pauling Institute. Oregon State University. http://lpi.orst.edu/

4. Institute of Medicine. Food and Nutrition Board. *Dietary Reference Intakes for Thiamin, Riboflavin, Niacin, Vitamin B6, Folate, Vitamin B12, Pantothenic Acid, Biotin, and Choline.* 1998.

VITAMIN B6 (PYRIDOXINE)

WHAT IS VITAMIN B6?

Vitamin B6, also called **pyridoxine**, is a water-soluble vitamin that helps keep the body physically and mentally healthy. It is a critical vitamin involved in **anabolic** and **catabolic** reactions of amino acids, and the neurotransmitters **serotonin**, **melatonin**, and **dopamine**.

WHAT DOES VITAMIN B6 DO?

Vitamin B6 can help prevent dry skin conditions such as dandruff, eczema, and psoriasis. Vitamin B6 helps the immune system, increases the production of red blood cells and the growth of new cells, and helps balance **sodium** and **potassium**. Vitamin B6 is used in the metabolism of proteins, fats, and carbohydrates.

Vitamin B6 plays a role in determining mood and behavior because of its role in **neurotransmitter formation**. Vitamin B6 works in conjunction with Vitamin B12 and folic acid to reduce the amount of **homocysteine** produced in the body.

WHAT ARE THE EFFECTS OF VITAMIN B6 DEFICIENCY?

Although Vitamin B6 deficiency is not common in North America, individuals with high alcohol consumption have an increased risk of developing a deficiency.

Individuals with Vitamin B6 deficiency may feel nervous and irritable, have trouble sleeping, and may feel weak. Dermatitis, acne, ridged nails, and a swollen tongue are common symptoms. Individuals with Vitamin B6 deficiency may develop allergies or asthma, and may be at increased risk of developing arthritis and osteoporosis.

In pregnant women, vitamin B6 deficiency may cause nausea, vomiting, cramps, and pain in the arms and legs. Women who have a Vitamin B6 deficiency and are taking birth control pills or hormone replacement therapy may experience mood swings, depression, and a low sexual drive.

Vitamin B6 is required to metabolize **homocysteine** to **cysteine**. As we'll discuss later, when we talk about **Vitamin B12,** elevated homocysteine blood levels are a significant risk factor for the development of cardiovascular disease. Homocysteine is metabolized to methionine via one pathway and to cysteine via another. Thus lowered serum concentrations of Vitamin B6 can hamper the conversion of homocysteine to cysteine[3].

IS VITAMIN B6 SUPPLEMENTATION RECOMMENDED?

Some studies have shown that Vitamin B6 supplements improve **glucose tolerance** in women suffering from **gestational diabetes**[2]. Vitamin B6 supplementation has also been found to reduce the number of asthma attacks in people with mild to moderate asthma[1].

VITAMIN B6 TOXICITY

Toxic intakes of Vitamin B6 are extremely rare, but if they occur they can cause sensory nerve impairment and loss of muscular control. Doses of 2,000 mg or more per day—an extremely large dose—of B6 supplements can cause neurological damage.

SOURCE S OF VITAMIN B6

Meats—especially beef and organ meats, whole grains and whole grain products, brewer's yeast, wheat germ, pecans, bananas, green leafy vegetables, eggs, milk, and cabbage are all good sources of Vitamin B6.

Food sources for Vitamin B6

Food	Amount (mg)
Beef liver, 3 oz.	1.22
Oatmeal, ¾ c.	0.74
Banana	0.66
Chicken, 3 oz.	0.51
Mashed potatoes, 1 c.	0.49
Brewer's yeast, 1 T.	0.40
Baked halibut, 3.5 oz.	0.34
Pork chop, 3 oz.	0.30
Cooked brown rice, 1 c.	0.28
Hamburger, 3 oz.	0.23

VITAMIN B6 REQUIREMENTS

It is recommended that 0.016 mg of Vitamin B6 be consumed for every gram of protein consumed. Thus an average female should consume 1.6

mg of Vitamin B6 per day while a man should consume 2 mg of Vitamin B6 per day. High consumption of fat or protein raises requirements for Vitamin B6, because it plays a role in the metabolism of fats and proteins.

People with asthma often have Vitamin B6 deficiencies. This could be due to the medicines used to treat asthma, which affect Vitamin B6 absorption[4].

REFERENCES

1. Sur, S., Camara, M., Buchmeier, A., et al. Double-blind trial of pyridoxine (Vitamin B6) in the treatment of steroid-dependent asthma. *Ann Allerg* 1993;70:141-52.

2. Ambrosch, A., Dierkes, J., Lobman, R., Kuhne, W., Konig, W., Luley, C. & Lehnert, H. Relationship between homocysteinaemia and diabetic neuropathy in patients with type II diabetes melliteis. *Diabetic Medicine* 2001; 18(3): 185-192.

3. El-Khairy, L., Ueland, P. M., Refsum, H., Graham, I. & Vollset, S. E. Plasma roral cysteine as a risk factor for vascular disease: The European concerted action project. *Circulation* 2001; 103: 2544-2549.

4. Kaslow, J. E. Double blind trial of pyridoxine (Vitamin B6) in the treatment of steroid dependent asthma. *Annals of Allergy* 1993; 70(2): 147-152.

BIBLIOGRAPHY

1. Linus Pauling Institute. Oregon State University. http://lpi.orst.edu/

2. Ziegler, E. & Filer, L. J. (Eds.) *Present Knowledge in Nutrition*, 7th edition. Washington DC: ILSI Press, 1996.

VITAMIN B12 (COBALAMIN)

WHAT IS VITAMIN B12?

Vitamin B12, also known as **cobalamin**, is a water-soluble vitamin. It is a crystalline compound containing one atom of cobalt per molecule of cobalamin.

In 1948, Vitamin B12 was isolated from a liver extract and was found to be the active component in liver that prevented and cured **pernicious anemia**—a disease that until 1926 had nearly always been fatal. (In 1926, Minot and Murphy discovered a cure for pernicious anemia[2]. Murphy won the Nobel Prize for the discovery.)

WHAT DOES VITAMIN B12 DO?

In conjunction with **folic acid**, Vitamin B12 is involved in the synthesis of red blood cells and DNA. Vitamin B12 is also involved in nerve transmission and is used in the synthesis of the **myelin sheath**, which speeds the transmission of nerve signals.

Vitamin B12 is also an important cofactor for metabolic reactions called **demethylation** and **methylation** reactions. Vitamin B12 is intrinsic to the synthesis of **methionine**. Methionine plays an important role in methylation, particularly for DNA and RNA synthesis. Proper methylation of DNA and RNA is critical for the control of normal cell growth and reproduction[3].

In addition, methionine is synthesized from **homocysteine**. Recent research has shown that high levels of homocysteine can lead to increased risk of cardiovascular disease[4].

ARE VITAMIN B12 SUPPLEMENTS NECESSARY?

Vitamin B12 supplements are not normally necessary if the diet includes daily servings of meat, milk, and other dairy products.

Vegetarians are likely to need a Vitamin B12 supplement. The **elderly** may also need a Vitamin B12 supplement because the body may lose the capacity to absorb Vitamin B12 as it ages. This typically results from age-related disorders of the gastrointestinal tract such as achlorhydria, a condition characterized by low stomach acid secretion.

VITAMIN B12 DEFICIENCY

Vitamin B12 deficiency can be caused by inadequate dietary intake. However, the majority of Vitamin B12 deficiencies occur in people who do not secrete enough **intrinsic factor.** Intrinsic factor, secreted by specialized cells in the lining of the stomach, aids the body in absorption of Vitamin B12.

Conditions such as celiac disease or surgical resection of the intestine may also affect the absorption of Vitamin B12. Vitamin B12 is different than other water-soluble vitamins in that it is stored in the liver, kidney, and other tissues. The body's use of stored Vitamin B12 may delay symptoms of dietary deficiency for 5 to 6 years.

Pernicious anemia is the disease most commonly associated with Vitamin B12 deficiency. Anemia is a disease characterized by an abnormally low level of red blood cells in the blood. As a result, not enough oxygen is delivered to the tissues and organs. Pernicious means "having the ability to kill, destroy, or cause great harm." This disease is so named because, if not treated, it is almost always fatal.

However, the first areas of the body to be affected by a deficiency are the **brain** and the **nervous system**. Numbness, tingling, or burning sensations may occur as a result of impaired nerve function caused by Vitamin B12 deficiency. Vitamin B12 deficiency may also impair mental function. When this occurs in elderly people, it is often mistaken for Alzheimer's disease. Deficiency can also cause the tongue to become smooth, red, and beefy.

SOURCES OF VITAMIN B12

Vitamin B12 is not found in plants, but comes from animals and animal products. The best source of Vitamin B12 is liver and kidneys, but other meats, and eggs, fish, milk, and milk products are also good sources.

Because only small amounts Vitamin B12 are available from foods, the body has an adaptive mechanism to enhance absorption. Specialized cells in the stomach lining secrete a substance called **intrinsic factor** that binds with the vitamin and increases its absorption by the small intestine.

RDA for Vitamin B12[5]

Children	RDA (μg)
Under 6 months	0.4
6 months–1 year	0.5
1–3 years	0.9
4–8 years	1.2
Adolescents	
9–13 years	.8
14–18 years	2.4
Adults	
19 years and older	2.4
Pregnant women	2.6
Breastfeeding women	2.8

REFERENCES

1. Combe, J. S. History of a case of anaemia. *Trans Med Chirurg Soc* Edinburgh 1:194-204. 1824

2. Minot, G. R. & Murphy, W. P. Treatment of pernicious anemia by special diet. *JAMA* 87:470-476. 1926

3. Weinstein, S. J., Ziegler, R. G., Selhub, J., Fears, T. R., Strickler, H. D., Brinton, L. A. et al. Elevated serum homocysteine levels and increased risk of invasive cervical cancer in US women. *Cancer Causes and Control* 2001; 12(4): 317-324.

4. Ozkan,Y., Ozkan, E. & Simsek, B. Plasma total homocysteine and cysteine levels are cardiovascular risk factors in coronary heart disease. *Int J of Cardio*, 2002; 82(3): 269-277.

5. Institute of Medicine. Food and Nutrition Board. *Dietary Reference Intakes for Thiamin, Riboflavin, Niacin, Vitamin B6, Folate, Vitamin B12, Pantothenic Acid, Biotin, and Choline.* 1998.

BIBLIOGRAPHY

1. Linus Pauling Institute. Oregon State University. http://lpi.orst.edu/

2. Ziegler, E. & Filer, L. J. (Eds.) *Present Knowledge in Nutrition*, 7th edition. Washington DC: ILSI Press, 1996.

VITAMIN C (ASCORBIC ACID)

The benefits of Vitamin C consumption have been documented for many years. English sailors of the eighteenth century earned the nickname "limeys" due to their habit of eating limes every day to prevent scurvy. Although the use of limes to prevent scurvy was a common practice, the compound responsible was not isolated until 1928 when the Nobel Prize winning Hungarian scientist, Dr. Szent-Gyorgyi, isolated Vitamin C. Since then, there has been much research conducted on Vitamin C and its role in the body.

WHY DOES THE BODY NEED VITAMIN C?

Vitamin C, or **ascorbic acid**, is a water soluble vitamin best known for its role in maintaining a healthy **immune system**. The immune system is the body's defense mechanism against invading pathogens, infections and chronic diseases. Vitamin C is used in the production of white blood cells, T cells, and macrophages, which are components of the immune system.

Vitamin C is also an **antioxidant**[1,8,9]. Antioxidants act by scavenging free radical species. Although free radicals are a product of normal metabolism, they are also involved in the initiation, promotion, and progression of chronic diseases such as cancer and cardiovascular diseases. Thus, it is important to include adequate antioxidants in your diet to prevent the oxidative damage caused by free radicals.

In addition, the body's first line of defense against any sort of foreign insult (pathogenic bacteria, virus, environmental factors, etc.) is the skin and mucosal linings of the alimentary and respiratory tracts. The body requires Vitamin C for the synthesis of key structural proteins such as **collagen**.

DAILY REQUIREMENTS OF VITAMIN C

The Vitamin C Recommended Daily Allowance is 60 mg per day, the amount required to prevent scurvy. A diet containing five servings daily of fresh fruits and vegetables provides about 200 mg of Vitamin C. In the United States, less than 15% of children and adults consume the recommended amount[2,4].

ARE VITAMIN C SUPPLEMENTS NECESSARY?

The human body lacks the ability to synthesize Vitamin C, thus a dietary source is required. Because Vitamin C is water-soluble, Vitamin C that is not utilized is excreted in the urine.

Individuals with certain medical conditions may benefit from a Vitamin C supplement. Research has shown that levels of Vitamin C in the blood are much lower in people who have asthma, arthritis, cancer, diabetes, or heart disease than in healthy people[5].

WHAT ARE THE EFFECTS OF VITAMIN C DEFICIENCY?

Fatigue is one of the first symptoms of Vitamin C deficiency. Vitamin C deficiency weakens the **immune system**, which reduces the body's ability to prevent and fight disease and infection. In severe Vitamin C deficiency, **scurvy**, a disease that causes capillary hemorrhaging and degeneration of collagenous tissue, may result.

WHAT ARE THE RISKS OF TAKING LARGE DOSES OF VITAMIN C?

Megadosing of Vitamin C can interfere with absorption of other nutrients. In diabetics, certain kidney disorders may cause very large doses of Vitamin C to have toxic effects[7]. Although doses in excess of 500 mg per day in healthy people[2,4] generally do not have any negative effects, individuals who are prone to **oxalate kidney stones** may have increased incidence of stone formation at high dosages[10,11].

SOURCES OF VITAMIN C

The most common sources of Vitamin C are fruits and vegetables, especially citrus fruits and tomatoes. Juices fortified with Vitamin C are an excellent source as well. Many cereals are also fortified with Vitamin C.

Special techniques used in food preparation and storage may prevent the loss of Vitamin C. Because Vitamin C is unstable when heated, uncooked fruits and vegetables contain the highest levels of the vitamin. Remember that vitamin C is water soluble; when cooking fruits and vegetables rich in Vitamin C, cook them in as little water a possible to prevent the Vitamin C from leaching out of the food and into the cooking water.

Sources of vitamin C

Food	Serving size	% of RDA for adults and children over 4 years old
Fortified cereals	1 oz.	25–39
Raw apple	1 medium	10–24
Baked, unsweetened apple	1 medium	10–24
Fortified apple juice	¾ c.	40 or higher
Raw banana	1 medium	10–24
Raw grapefruit	½ medium	40 or higher
Grapefruit juice	¾ c.	40 or higher
Raw honeydew melon	¾ c.	40 or higher
Raw kiwi fruit	1 medium	40 or higher
Mandarin orange	½ c.	40 or higher
Raw mango	½ medium	40 or higher
Raw nectarine	1 medium	10–24

Food	Serving size	% of RDA for adults and children over 4 years old
Raw orange	1 medium	40 or higher
Orange juice	¾ c.	40 or higher
Raw papaya	¼ c.	40 or higher
Raw peach	1 medium	10–24
Raw pear	1 medium	10–24
Pineapple	½ c.	10–24
Unsweetened canned pineapple juice	¾ c.	25–39
Raw plum	1 medium	10–24
Raspberries	½ c.	25–39
Strawberries	½ c.	40 or higher
Raw tangerine	1 medium	40 or higher
Cooked artichoke	1 medium	10–24
Cooked asparagus	½ c.	40 or higher
Cooked green or yellow beans	½ c.	10–24
Cooked lima beans	½ c.	10–24
Bean sprouts	½ c.	10–24
Broccoli	½ c.	40 or higher
Cooked Brussels sprouts	½ c.	40 or higher
Cooked Chinese cabbage	½ c.	25–39
Green cabbage	½ c.	25–39
Red cabbage	½ c.	40 or higher
Cauliflower	½ c.	40 or higher
Collards	½ c.	10–24
Cooked kale	½ c.	40 or higher
Cooked kohlrabi	½ c.	40 or higher
Cooked okra	½ c.	10–24
Cooked onion	1 large	10–24

Food	Serving size	% of RDA for adults and children over 4 years old
Raw onion	1 medium	10–24
Cooked green peas	½ c.	10–24
Sweet, green, or red pepper	½ c.	40 or higher
Boiled plantain	1 medium	40 or higher
Baked potato with skin	1 medium	25–39
Boiled potato with skin	1 medium	25–39
Raw radishes	6 large	10–24
Cooked rutabagas	½ c.	25–39
Snow peas	½ c.	40 or higher
Sweet potato	1 medium	40 or higher
Canned tomatoes	½ c.	40 or higher
Cooked tomatoes	½ c.	25–39
Raw tomato	1 medium	25–39
Tomato juice	¾ c.	40 or higher
Cooked turnips	½ c.	10–24
Beef or pork liver	3 oz.	25–39
Chicken liver	½ c.	10–24
Clams	3 oz.	10–24
Mussels	3 oz.	10–24

REFERENCES

1. Sies, H. & Stahl, W. Vitamins E and C, beta-carotene, and other carotenoids as antioxidants. *Am J Clin Nutr*, 1995; 62: 1315S-21S.

2. Levine, M. New concepts in the biology and biochemistry of ascorbic acid. *New Engl J Med*, 1986; 314:892-902 .

3. Pauling, L. *Vitamin C and the Common Cold*. Freeman, San Francisco, CA, 1970.

4. Levine, M., Conry-Cantilena, C., Wang, Y. et al. Vitamin C pharmacokinetics in healthy volunteers: Evidence for a recommended dietary allowance. *Proceedings of the National Academy of Sciences USA*, Vol. 93, No. 8, April 16, 1996, pp. 3704-09.

5. Douglas, R. M., Chalker, E. B. & Tracey, B. Vitamin C for preventing and treating the common cold. *Cochrane Database Syst Rev* 200:2 CD 00980

6. Rath, M. Eradicating Heart Disease. *Health Now*, San Francisco, CA, 1993.

7. Goldberg, B. *Alternative Medicine: The Definitive Guide*. Future Medicine Publishing. Puyallup, WA, 1993.

8. Block, G. The data support a role for antioxidants in reducing cancer risk. *Nutrition Reviews*, 50(7):1992; 207-13.

9. Frei, B. Reactive oxygen species and antioxidant vitamins: mechanisms of action. *American Journal of Medicine*, 97:3A1994: 5S-13S.

10. Wandzilak, T. R., D'andre, S. D., Davis, P. A. & Williams, H. E. Effects of high dose vitamin C on urinary oxalate levels. *J Urol.* 1995;151(4):834-837.

11. Urivetzky, M., Kessaris, D. & Smith, A. D. Ascorbic Acid overdosing: a risk factor for calcium oxalate nephrolithiasis. *J Urol.* 1992;147(5):1215-1218.

12. Anderu, C., Patulny, R. V., Sander, B. H. & Douglas, R. M. Megadose Vitamin C in treatment of the common cold: a randomized controlled trial. *Med J Aust* 2001:175:359-362.

BIBLIOGRAPHY

1. Linus Pauling Institute. Oregon State University. http://lpi.orst.edu/

2. Ziegler, E. & Filer, L. J. (Eds.) *Present Knowledge in Nutrition*, 7th edition. Washington DC: ILSI Press, 1996.

VITAMIN D

WHAT IS VITAMIN D?

Vitamin D is a very unique fat-soluble vitamin. Although it can be obtained from food sources, with adequate exposure to sun the body can synthesize Vitamin D from 7-dehydrocholesterol present in the skin[2]. Unfortunately, the synthesis of Vitamin D in not as efficient for people who have higher concentrations of pigment in their skin (e.g. African Americans).

Sunscreen use can also reduce the formation of vitamin D in the skin.

WHAT DOES VITAMIN D DO?

In the early twentieth century, the disease **rickets** afflicted many children. Rickets is characterized by the improper development of the bones. Individuals with rickets had soft pliable bones that resulted in deformities such as bow legs and knock-knees. It was discovered that rickets was caused by a Vitamin D deficiency.

The main purpose of Vitamin D is to regulate the levels of **calcium** and **phosphorous**, minerals that are essential for maintaining bone health and muscle function. Vitamin D from the diet is metabolized by the body to form 1, 25-dihydroxy vitamin D (**calcitriol**) which functions as a hormone in the body to regulate calcium and phosphorous homeostasis in the body.

Calcitriol works in conjunction with parathyroid hormone to maintain calcium homeostasis. When there is a decrease in blood concentration of calcium, the parathyroid gland secretes parathyroid hormone which in turn triggers formation of calcitriol. Calcitriol then stimulates the uptake of calcium from the intestine and increases the release of calcium and phosphorous from bone, resulting in an increased serum concentration of these two minerals.

SOURCES OF VITAMIN D

Milk is an excellent source of Vitamin D. Vitamin D is present in very few other foods, unless they have been fortified or enriched.

WHEN CAN VITAMIN D DEFICIENCY OCCUR?

Vitamin D deficiency can occur with inadequate consumption of foods containing Vitamin D or exposure to the sun.

As we have already discussed, Vitamin D deficiency can cause rickets in children. Adults who do not get enough Vitamin D or calcium and phosphorous can also develop a bone disorder called **osteomalacia**. Osteomalacia is characterized by a softening of the bones accompanied by weakness, fracture, pain, anorexia, and weight loss. Osteomalacia is usually accompanied by some underlying disorder. Treatment for osteomalacia involves the administration of vitamins and minerals as well as treatment for the underlying disorder[1].

VITAMIN D REQUIREMENTS

Adequate intakes for Vitamin D[3]

Age and gender	Adequate intake level (µg/day)
Men 19–50	5
Women 19–50	5
Men 51–69	10
Women 51–69	10
Men 70 and older	15
Women 70 and older	15

Source: The Food and Nutrition Board

REFERENCES

1. Goldring, S. R., Krane, S. & Avioli, L. V. Disorders of calcification: Osteomalacia and rickets. LJ D, (Ed.) *Endocrinology*. 3rd ed. Philadelphia: WB Saunders, 1995:pp.1204-1227.

2. Holick, M. F. McCollum Award Lecture, 1994: Vitamin D: New horizons for the 21st century. *Am J Clin Nutr* 1994;60:619-630.

3. Institute of Medicine. Food and Nutrition board. *Dietary Reference intake for Calcium, Phosphorus, Magnesium, Vitamin D, and Fluoride.* 1997.

BIBLIOGRAPHY

1. DeLuca, H. F. & Zierold, C. Mechanisms and functions of vitamin D. *Nutr Rev* 1998;56:S4-10.

2. Groff, J.L., Gropper, S. S. & Hunt, S. M. The fat soluble vitamins. In: *Advanced Nutrition and Human Metabolism.* 2nd Edition. 1995. West Publishing Company. St Paul, MN. pp.298-306.

3. Linus Pauling Institute. Oregon State University. http://lpi.orst.edu/

4. Ziegler, E. & Filer, L. J. (Eds.) *Present Knowledge in Nutrition,* 7th edition. Washington DC: ILSI Press, 1996.

VITAMIN E

Vitamin E is a fat-soluble vitamin that has attracted media and research attention. Vitamin E has been widely popularized due to its incorporation into many personal care products, in particular anti-wrinkle lotions and creams. Foods containing Vitamin E are thought to provide beneficial antioxidant and cardiovascular properties[3]. It is not clear whether Vitamin E supplements provide similar benefits.

WHAT IS VITAMIN E?

Vitamin E is a fat-soluble vitamin found in eight different forms, each one varying in its degree of potency[1]. These eight different forms of Vitamin E are grouped into two classes: the **tocols** and the **tocotrienols**. All the forms of Vitamin E have the same basic structure with just slightly differing side chains, giving each a different degree of biological activity. Of the different forms of Vitamin E, **Alpha tocopherol**, **Beta tocopherol**, and **Alpha tocotrienol** have the most biological activity.

WHEN CAN VITAMIN E DEFICIENCY OCCUR?

Development of a Vitamin E deficiency is rare. However, it may occur in individuals who have problems in fat metabolism and absorption, such as cystic fibrosis, resection of the stomach, inflammatory bowel disease, and disorders where the bile production is affected. Individuals with abetalipoproteinemia, a rare inherited disorder characterized by impaired fat absorption, are predisposed to Vitamin E deficiency. Some symptoms of Vitamin E deficiency include muscle weakness, degenerative neurological problems, problems with coordination, and retinal degeneration. Supplementation is recommended for individuals with problems in fat metabolism and absorption.

RECOMMENDED DIETARY ALLOWANCE FOR VITAMIN E

The RDA for Vitamin E is 10 mg. The RDA for Vitamin E was originally 8mg/day for women and 10mg/day for men, but this was updated by the Food and Nutrition Board of the Institute of Medicine in 2000[2].

The update was recommended in response to tests carried out in the 1950s in men who were put on diets low in Vitamin E. During these tests, hydrogen peroxide was added to blood samples and the resulting hemolysis was taken as an indicator of Vitamin E deficiency. This was considered relevant clinical testing due to the fact that hemolysis was also reported in children with severe Vitamin E deficiency.

RDA for Vitamin E

Life Stage	Age	Males (mg/day)	Females (mg/day)
Infants	0–6 months	3	3
Infants	7–12 months	4	4
Children	1–3 years	5	5
Children	4–6 years	6	6

Life Stage	Age	Males (mg/day)	Females (mg/day)
Children	7–10 years	7	7
Adolescents	11–14 years	8	8
Adults	15 and older	10	8
Pregnancy	All ages	N/A	+2
Breastfeeding	All ages	N/A	+3

SOURCES OF VITAMIN E

Vitamin E is abundantly available in many foods. However, it is more abundant in foods of plant rather than animal origin. Leafy green vegetables, including the green stalks and stems, are good sources of Vitamin E. Nut, seed, and vegetable oils are good sources of Vitamin E. In addition, many breakfast cereals are fortified with Vitamin E.

Food Sources for vitamin E

Food	IU	% DV
Wheat germ oil, 1 T.	26.2	90
Almonds, dry roasted, 1 oz.	7.5	25
Safflower oil, 1 T.	4.7	15
Corn oil, 1 T.	2.9	10
Soybean oil	1 T.	2.5
Turnip greens, frozen, boiled, 1.5 c.	2.4	8
Mango, raw, without refuse, 1 fruit	2.3	8
Peanuts, dry roasted, 1 oz.	2.1	8
Mixed nuts w/peanuts, oil roasted, 1 oz .	1.7	6
Mayonnaise, made w/ soybean oil, 1 T.	1.6	6
Broccoli, froze, chopped, boiled, 1.5 c.	1.5	6
Dandelion greens, boiled, 1.5 c.	1.3	4
Pistachio nuts, dry roasted, 1 oz.	1.2	4
Spinach, frozen, boiled, 1.5 c.	0.85	2
Kiwi, 1 medium fruit	0.85	2

DV is the daily value, based on the Recommended Dietary Allowances (RDA). This number allows one to determine how much of a certain nutrient is available in food. Daily values are determined based on a 2000 calories per day diet. Daily values will differ with different diets.

REFERENCES

1. Traber, M. G. Vitamin E. In M. Shils, J. A. Olson, M. Shike & A. C. Ross (Eds.) *Nutrition in Health and Disease.* 9th ed. Baltimore: Williams & Wilkins; 1999:pp. 347-362.

2. Food and Nutrition Board, Institute of Medicine. *Vitamin E. Dietary reference intakes for vitamin C, vitamin E, selenium, and cartenoids.* Washington D.C.: National Academy Press; 2000:95-185.

3. Keaney, J. F., Simon, D. I. & Freedman, J. E. Vitamin E and vascular homeostasis: implications for atherosclerosis. *FASEB J.* 1999; 13: 965-976.

VITAMIN K

WHAT IS VITAMIN K?

Vitamin K is a fat-soluble vitamin found in nature and made in the body. Vitamin K can be found in three forms: **K_1, K_2 and K_3**.

- Vitamin K_1 is the form naturally present in food.
- K_2 is the form synthesized by the bacteria in the body.
- K_3 is a synthetic form with twice the biological activity of K_1 and K_2.

WHAT DOES VITAMIN K DO?

Vitamin K is involved in two major processes in the body. Vitamin K plays an essential role in the body's **blood clotting** process. Vitamin K is necessary for the synthesis of several clotting factors, such as **prothrombin**, a precursor to **thrombin**, which is an essential component in the clotting cascade.

In addition to its role in synthesis, Vitamin K also acts as a coenzyme to a Vitamin K dependent **carboxylase**. Carboxylase plays an essential role in the cycling of the clotting factor **prothrombin**.

Recent research has suggested a possible role for Vitamin K in **bone strength**. Disorders associated with Vitamin K deficiency show an increased incidence of hip fractures[2] and osteoporosis. Vitamin K is a cofactor in the carboxylation of the protein **osteocalcin**. Osteocalcin is a component of the bone matrix, thus is thought to contribute to bone strength. In deficiency syndromes, it is suspected that there is an inadequate carboxylation of this essential bone matrix protein[1].

VITAMIN K REQUIREMENTS

The adequate intake levels for Vitamin K are 120 µg per day for men 19 years and older, and 90 µg per day for women 19 years and older[3]. Most diets provide sufficient Vitamin K, so an official RDA (Recommended Daily Allowance) has not been established.

VITAMIN K DEFICIENCY

Although Vitamin K deficiency is relatively rare, it can occur in individuals who have problems with fat absorption, who have liver damage, or who are on anticoagulant therapy. However, dietary deficiencies of Vitamin K are very rare, because Vitamin K is found in so many foods.

Individuals with Vitamin K deficiency may show impaired blood clotting, resulting in easy bruising and excessive bleeding ranging from frequent and lasting nosebleeds to very heavy menstrual bleeding.

TOXICITY

There is no known toxicity level for vitamin K. Because the body is extremely efficient, both in storing and in eliminating excess amounts of the natural forms of Vitamin K, toxicity from natural Vitamin K is rare.
Synthetic forms of Vitamin K are much more difficult to store and eliminate, so slight toxic levels may build in the bloodstream. However

this type of toxicity is also rare. Symptoms of a Vitamin K overdose include flushing, sweating, or chest tightness.

SOURCES OF VITAMIN K

Vitamin K is present in many foods of plant or animal origin. Green leafy vegetables, such as spinach and broccoli, are excellent sources of the vitamin. It is also present in alfalfa, kelp, molasses, and in many edible oils such as safflower oil. Vitamin K is found in significant proportions in liver, but milk, yogurt, egg yolks, and fish oils are also good sources.

Although the bacteria in the intestines can produce vitamin K, it has not been determined whether the body can absorb the vitamin in this from in significant amounts.

Food sources of Vitamin K[4]

Food	Amount of vitamin K (μg)
Olive oil, 1 T.	6.6
Soybean oil, 1 T.	26.1
Canola oil, 1 T.	19.7
Mayonnaise, 1 T.	11.9
Cooked broccoli, 1 c.	420
Raw kale, 1 c.	547
Raw spinach, 1 c.	120
Raw lettuce, 1 c.	118
Raw watercress, 1 c.	85
Raw parsley, 1 c.	324

REFERENCES

1. Zitterman, A. Effects of vitamin K on calcium and bone metabolism. *Curr Opin Clin Nutr Met Care* 2001; 4:483-487.

2. Caraballo, P. J., Gabriel, S. E., Castro, M. R. et al. Changes in bone density after exposure to oral anticoagulants: a meta-analysis. *Osteoporos Int.* 1999;9:441-448.

3. Institute of Medicine. Food and nutrition Board. *Dietary reference intakes for Vitamin A, Vitamin K, Arsenic, Boron, Chromium, Copper, Iodine, Iron, Manganese, Molybdenum, Nickel, Silicon, Vanadium and Zinc.* 2002

4. Linus Pauling Institute. Oregon State University. http://lpi.orst.edu/

BIBLIOGRAPHY

1. Linus Pauling Institute. Oregon State University. http://lpi.orst.edu/

2. Ziegler, E. & Filer, L. J. (Eds.) *Present Knowledge in Nutrition,* 7th edition. Washington DC: ILSI Press, 1996.

MINERALS

CALCIUM

WHAT IS CALCIUM?

Most calcium in the body is used to build and maintain bones and teeth. In fact, calcium is the mineral that bones and teeth are primarily composed of. The Food and Drug Administration (FDA) has approved the use of health claims in food labeling that link calcium with osteoporosis prevention.

Small amounts of calcium are also used in the cells of soft tissue and circulate in the blood and extracellular fluid. This calcium is important for proper nerve transmission, muscle contraction, blood clotting, and as many other bodily functions.

CALCIUM REQUIREMENTS

Calcium requirements vary significantly with age and gender. The table below provides a guide. Females need higher levels of calcium at certain stages of their lives, as is explained in the following paragraphs.

According to Abrams[2,] a higher bone deposition rate 4 to 5 years after **menarche** (that is, by age 15 to 16 in most girls), resulting in peak bone mass. Girls require increased calcium during this stage to build their bones during the rapid growth often experienced at this stage. A small increase in bone density also occurs in women between the ages of 18 and 30. During the second decade in a girl's life, there is a decrease in calcium requirements because the bone mass becomes stable. This remains the case until menopause[3].

Older women experience substantial bone loss during early menopause, beginning within the first year and ending within 6 years. During this period the bone resorption rate is substantially higher than the bone formation rate. Calcium requirements are higher for women during this period because the body less effectively absorbs dietary calcium, due to a decrease in estrogen. This bone resorption to formation ratio eventually stabilizes[1].

Basic calcium requirements[5]

Age	Amount of calcium needed (mg/day)
0–6 months	210
6–12 months	270
1–3 years	500
4–8 years	800
9–18 years	1300
19–50 years	1000
51 years or older	1200

CALCIUM DEFICIENCY

A lifelong deficiency in calcium can result in a skeletal disorder called **osteoporosis**. The link between calcium deficiency and osteoporosis has been well established. Osteoporosis is characterized by decreased bone density and subsequent loss of bone strength. Individuals with osteoporosis are at risk for bone fractures and breaks, commonly in the hip, pelvis, vertebrae, and ribs.

Some studies have found low calcium levels to be associated with increased risk of **hypertension**.

WHAT ABOUT TOO MUCH CALCIUM?

In individuals with a history of **oxalate kidney stones**, too much calcium can increase the frequency of occurrence. Too much calcium can also result in **hypercalcemia**, a condition characterized by abnormally high concentrations of calcium compounds in the blood stream.

SOURCES OF CALCIUM

Calcium is present in dairy products and in many foods of plant origin. Dark, leafy green vegetables such as broccoli, kale, and rhubarb are very high in calcium.

It is also important to consider the **bioavailability** of calcium and the presence of factors that might hinder calcium absorption. Although some vegetables are very high in calcium, the calcium present is not bioavailable (readily absorbed), and the body therefore is unable to utilize it. In other vegetables, such as rhubarb, the high calcium content is countered by the presence of **oxalic acid**, which binds the calcium and renders it unusable by the body.

Food sources of calcium

Food	Amount of calcium (mg)
Grains	
Cooked brown rice, 1 c.	20
Corn bread, 2 oz.	133
Corn tortilla	42
English muffin	92
Pita bread, 1 piece	18
Wheat bread, 1 slice	18
All-purpose wheat flour, 1 c.	22
Whole wheat flour, 1 c.	40
Fruits	
Apple, 1 medium	10
Banana, 1 medium	7

Food	Amount of calcium (mg)
Dried figs, 10	269
Naval orange, 1 medium	56
Pear, 1 medium	19
Raisins, 2/3 c.	53
Vegetables	
Broccoli, 1 c.	94
Brussels sprouts, 1 c.	56
Butternut squash, 1 c.	84
Carrots, 2 medium	38
Cauliflower, 1 c.	34
Celery, 1 c.	64
Collards, 1 c.	348
Kale, 1 c.	94
Onions, 1 c.	46
Potato, 1 medium	20
Romaine lettuce, 1 c.	20
Sweet potato, 1 c.	70
Legumes	
Chick peas, 1 c.	78
Green beans, 1 c.	58
Green peas, 1 c.	44
Kidney beans, 1 c.	50
Lentils, 1 c.	37
Lima beans, 1 c.	32
Soybeans, 1 c.	175
Tofu, ½ c.	258

REFERENCES

1. Abrams, S. A. Calcium turnover and nutrition through the life cycle, *Proc Nutr Soc.* 2001; 60: 283-289.

2. Abrams, S. A., O'Brien, K. O. & Stuff, J. E. Changes in calcium kinetics associated with menarche. *J Clin Endocrinology and Metabolism* 1996; 81; 2017-2020.

3. The North American Menopause Society. Consensus Opinion: The role of calcium in peri- and postmenopausal women: consensus opinion of The North American Menopause Society. *Menopause.* 2001; 8(2):84-95.

4. Goel, V., Ooraikul, B. & Baru, T. K. Effect of dietary rhubarb stalk fiber on the bioavailability of calcium in the rats. *Int JFood Sci Nutr.* 1996; 47(2):159-163

5. Institute of Medicine. Food and Nutrition board. *Dietary Reference intake for Calcium, Phosphorus, Magnesium, Vitamin D, and Fluoride.* 1997.

CHROMIUM

WHAT IS CHROMIUM?

Chromium III is a trace mineral found in soil and in the foods grown in soil, in most parts of the world. This form of chromium is essential to the body. It plays a role in **glucose metabolism** by enhancing the effects of insulin. Through its interaction with insulin, chromium indirectly affects how fat and protein are metabolized.

Chromium is marketed as a supplement for **diabetes** prevention, because of its influence on glucose metabolism[2,3]. However, scientific data in support of chromium's use in preventing diabetes is weak.

Food sources of Chromium[4]

Food	Amount of chromium (μg)
American cheese, 1 oz.	48
Peanut butter, 1 T.	41
Cooked spinach, 1 c.	36
Chicken breast, 3 oz.	22
Mushrooms, 1 c.	20
Wheat bread, 1 slice	16
Apple, 1 medium	15

Adequate Intakes

Age	AI (μg/day)
Infants	
0–6 months	0.2
7–12 months	5.5
Children	
1–8	11–15
Males	
9–13	25
14–50	35
Over 50	30
Females	
9–13	21
14–18	24
19–50	25
Over 50	20
Pregnant women	29–30
Breastfeeding women	44–45

REFERENCES

1. Fryzek, J. P., Mumma, M. T., McLaughlin, M. K., Henderson, B. E. & Blot, W. J. Cancer mortality in relation to environmental chromium exposure. *JOEM*. 2001;43(7): 635-640.

2. Linus Pauling Institute. http://lpi.orst.edu/index.html

3. Stoecker, B. Chromium. In E. E. Ziegler & L.J. Filer (Eds.) *Present Knowledge in Nutrition*, 7th Edition. Washington DC: ILSI Press, 1996:pp. 344-352.

4. US Department of Agriculture: Agricultural Research Service. *Nutrient Database for Standard Reference.*

5. Institute of Medicine. Food and Nutrition Board. *Dietary reference intakes for Vitamin A, Vitamin K, Arsenic, Boron, Chromium, Copper, Iodine, Iron, Manganese, Molybdenum, Nickel, Silicon, Canadium, and Zinc*. 2001.

COPPER

WHAT IS COPPER?

The trace element copper is essential for many functions in the body. Several enzymes involved in energy production, tissue formation, the metabolism of iron, and the synthesis and metabolism of neurotransmitters require copper to carry out their functions. The majority of copper in the body is found in the kidney, liver, brain, heart and bones[1,2,3].

COPPER REQUIREMENTS[4]

Age and gender	RDA (mg/day)
Children 0–3 years	0.4–1
Children 4–6 years	1–1.5
Children 7–10 years	1–2
Females 11 years and older	1.5–3
Males 11 years and older	1.5–2.5

SOURCES OF COPPER

Copper is found in seafood, legumes, cereal, nuts, vegetables and fruits, and muscle meats[2].

REFERENCES

1. Sandstead, H. H. Requirements and toxicity of essential trace elements, illustrated by zinc and copper. *Am J Clin Nutr* 1995;61(suppl):62S-64S.

2. Linus Pauling Institute. http://lpi.orst.edu/index.html

3. Linder, M. C. Copper. In E. E. Ziegler & L.J. Filer (Eds.) *Present Knowledge in Nutrition*, 7th Edition. Washington DC: ILSI Press, 1996: pp. 307-319.

4. Institute of Medicine. *Dietary reference intakes for vitamin A, vitamin K, boron, chromium, copper, iodine, iron, manganese, molybdenum, nickel, silicon, vanadium and zinc.* Washington, D.C.: National Academy Press. 2001.

FLUORIDE

WHAT IS FLUORIDE AND WHAT DOES IT DO?

The trace element fluoride is not considered to be an essential mineral, but it has been found to enhance dental health. Teeth are constantly exposed to various organic compounds that demineralize them. Fluoride, in conjunction with calcium, helps remineralize teeth.

FLUORIDE REQUIREMENTS

The Food and Nutrition Board has based adequate intake levels on levels adequate to prevent dental cavities.

Fluoride Requirements[3]

Age	Adequate intake level (mg/day)
0–6 months	0.01
6–12 months	0.5
1–3 years	0.7
4–8 years	1.0
9–13 years	2.0
14–18 years	3.0
Men 19 years and older	4.0
Women 19 years and older	3.0

FLUORIDE TOXICITY

Fluoride can be toxic in large concentrations (5 mg/kg of body weight). Symptoms of fluoride toxicity include nausea, vomiting, abdominal pain, sweating, diarrhea, and weakness.

Too much fluoride during childhood can result in **dental fluorosis**. Strictly speaking, this is not a toxic effect. Fluorosis is a mottled enamel caused by the incorporation of excess fluoride into tooth enamel during maturation. Fluorosis can be alleviated by cosmetic dental surgery[6].

SOURCES OF FLUORIDE

Fluoride is present in many foods in trace amounts, usually less than 0.05 mg per 100 g. The majority of dietary fluoride comes from drinking water. Many municipalities fluoridate their water. Municipalities in warmer climates add a lower concentration (0.6 ppm) of fluoride to the water because people tend to drink more water in the heat, while colder climates add more fluoride (1 ppm)[1].

Food sources of fluoride[2]

Food	Amount of fluoride (mg)
Tea, 100 ml.	0.1–0.6
Canned sardines with bones, 3.5 oz.	0.2–0.4
Fish, 3.5 oz.	0.01–0.17
Chicken, 3.5 oz.	0.06–0.10

REFERENCES

1. Record, S., Montgomery, D. F. & Milano, M. Practice Guidelines: Fluoride supplementation and caries prevention. *J of Pediatric Health Care.* 2000; 14(5):247-249.

2. Institute of Medicine, Food and Nutrition Board. *Dietary Reference Intakes: Calcium, Phosphorus, Magnesium, Vitamin D, and Fluoride.* Washington, DC: National Academy Press, 1997: pp. 288-313.

3. Linus Pauling Institute. Oregon State University. http://lpi.orst.edu/

BIBLIOGRAPHY

1. Ziegler, E. & Filer, L. J. (Eds.) *Present Knowledge in Nutrition,* 7th edition. Washington DC: ILSI Press, 1996.

IRON

WHAT DOES IRON DO?

Iron is an essential component of **hemoglobin** and **myoglobin**. These two compounds are involved in the transport and storage of oxygen within the body. Iron is also a component of **cytochromes**. Cytochromes are important to cellular energy generation and involved detoxification and metabolism of drugs in the body.

IRON REQUIREMENTS

The RDA for iron is 18 mg per day for women and 8 mg for men. Children require between 7 and 11 mg of iron per day[1].

IRON DEFICIENCY

Iron deficiency is one of the most common forms of nutritional deficiency in the world. It generally affects young women, babies, and in some cases vegetarians.

Iron deficiency occurs in young women due to a variety of factors including poor diet, rapid growth, and menstruation. Infants are the most susceptible to iron deficiency because milk is low in iron. Babies therefore need iron supplements.

Vegetarians are also at risk of developing iron deficiency because of inadequate dietary intake. Therefore vegetarians should also add iron supplements to their diet.

SOURCES OF IRON

Iron can be found in a variety of foods of plant or animal origin. However, the body's ability to utilize iron is affected by the form of iron, and other compounds present may either hinder or enhance iron absorption. Iron is present as **heme** or **non-heme** iron. Heme is much more easily absorbed by the body and is available in animal products such as lean red meats. Non-heme iron is abundantly available in plant products, dairy products and meat.

Eating foods high in **Vitamin C** enhances the body's absorption of iron. Other compounds that enhance iron absorption are **calcium**, **phosphate**, **bran**, **phytic acid** and **polyphenols**. Phytic acid is found in unprocessed whole grain products while polyphenols are found in tea and in some vegetables.

Food sources of Iron

Food	Amount of iron (mg)
Liver, 3.5 oz.	6.8
Braised beef, 3.5 oz.	6.8
T-bone steak, 3.5 oz.	3.2
Baked ground beef, 3.5 oz.	3.5
Cooked shrimp, 3 oz.	2.6
Roasted chicken with skin, 3.5 oz.	1.4
Roasted turkey without skin, 3.5 oz.	1.4
Boiled spinach, ½ c.	3.2
Boiled red kidney beans, 1 c.	5.2
Baked potato with skin, 1 large	2.7
Enriched white rice, 1 c.	1.9
Enriched egg noodles, 1 c.	2.5
Brown long grain rice, 1 c.	0.82
Boiled broccoli, ½ c.	0.66
Boiled egg, 1 large	0.59

REFERENCES

1. Institute of Medicine. Food and Nutrition Board. *Dietary Reference Intakes for Vitamin A, Vitamin K, Arsenic, Boron, Chromium, Copper, Iodine, Iron, Manganese, Molybdenum, Nickel, Silicon, Vanadium, and Zinc*. 2001.

MAGNESIUM

WHAT IS MAGNESIUM?

Magnesium is an essential cofactor in over 300 enzymatic reactions within the body. It plays a role in the proper function of nerves and muscles, including the heart. It also plays a structural role in maintaining bone strength and cellular integrity. Magnesium is required for several metabolic processes and during protein synthesis[2,6].

MAGNESIUM DEFICIENCY

Magnesium deficiency is rare, primarily because of the body's ability to conserve magnesium during times of shortage. However, clinical disorders may increase the risk of developing magnesium deficiency. These conditions include malabsorption syndromes such as Crohn's disease and ulcerative colitis. Other disorders that may affect magnesium status are renal dysfunction, endocrine disorders, and diabetes mellitus. It is also found in individuals who have chronic alcoholism and in cases of protein energy malnutrition (Kwashiorkor-Marasmus disease)[5,6].

Individuals with magnesium deficiency may experience symptoms including confusion, disorientation, depression, muscle cramps, loss of appetite, seizures, and arrhythmias[1].

MAGNESIUM REQUIREMENTS

RDAs for magnesium[3]

Age and gender	RDA for magnesium (mg/day)
Girls 14–18	360
Boys 14–18	410
Women 19–30	310
Men 19–30	400
Women 31 and older	320
Men 31 and older	420
Pregnant women	+ 40

SOURCES OF MAGNESIUM

Magnesium is present in many foods of animal and plant origin. The best sources of magnesium are green vegetables, nuts, seeds, and whole grains. Refined foods contain magnesium, but it is present in lower concentrations than in unprocessed foods.

Food sources of Magnesium[4]

Food	Amount of magnesium (mg)
100% bran, 2 T.	44
Dry roasted almonds, 1 oz.	86
Toasted wheat germ, 1 oz.	90
Pumpkin seeds, ½ oz.	75
Dry roasted cashews, 1 oz.	73
Cooked spinach, ½ c.	65
Bran flakes, ½ c.	60
Baked potato with skin, 1 medium	55
Baked potato without skin, 1 medium	40
Cooked soybeans, ½ c.	54
Dry roasted peanuts, 1 oz.	50
Peanut butter, 2 T.	50
Cooked lentils, ½ c.	35
Banana, 1 medium	34
Raw shrimp, 3 oz.	29
Whole wheat bread, 1 slice	24

REFERENCES

1. Rude, R. K. Magnesium deficiency: A cause of heterogeneous disease in humans. *J Bone Miner Res* 1998;13:749-758.

2. Wester PO. Magnesium. Am J Clin Nutr. 1987;45:1305-1312.

3. US Department of Agriculture Agricultural Research Service. *Nutrient Database for Standard Reference.*

4. Institute of Medicine. Food and Nutrition Board. *Dietary Reference Intakes: Calcium, Phosphorus, Magnesium, Vitamin D and Fluoride.* National Academy Press. Washington, DC, 1999.

5. Elisaf, M., Bairaktari, E., Kalaitzidis, R. & Siamopoulos, K. Hypomagnesemia in alcoholic patients. *Alcohol Clin Exp Res* 1998;22:244-246.

6. Shils, M. E. Magnesium. In E. E. Ziegler & L.J. Filer (Eds.) *Present Knowledge in Nutrition*, 7th Edition. Washington DC: ILSI Press, 1996. pp. 256-264.

MANGANESE

WHAT IS MANGANESE?

Manganese is an essential trace element. The term manganese is derived from the Greek term for magic. This is fitting because the body uses manganese in so many ways[1,2]. Manganese tends to be concentrated in the bones, pancreas, liver, and kidneys. There is a higher concentration of manganese in structures that are normally pigmented.

WHAT DOES MANGANESE DO?

Manganese functions as a cofactor or activator of many enzymes involved in the metabolism of carbohydrates, amino acids, and enzymes involved in the formation of cartilage and bone[3]. It is also involved in the body's energy production cycles and in many reactions[3]. Manganese also has antioxidant properties.

WHAT ARE THE EFFECTS OF MANGANESE DEFICIENCY?

Although manganese deficiency is well documented in animal studies, less is known about the occurrence and effect of manganese deficiency in humans. Manganese deficiencies generally occur in children on long-term total parenteral nutrition and has been noted in studies where manganese is purposely withheld[1,2].

Manganese deficiency exhibits different neurological symptoms in children than in adults. Children with severe manganese deficiency may experience convulsions, paralysis, or blindness, while adults show less severe symptoms such as dizziness, weakness, and hearing problems. It has also been reported that manganese deficiency may lead to dermatitis, decrease in clotting proteins, and in some cases hypercholesterolemia[3,4]. These symptoms can be alleviated by adding manganese supplements to the diet.

MANGANESE TOXICITY

There have been very few reported cases of manganese toxicity in healthy individuals.

WHERE DO WE GET MANGANESE?

The best food sources of manganese are whole grains and nuts. Other good sources are seeds, peas, beans, and leafy green vegetables. Vegetables have a higher content of manganese if they are grown in soil with high manganese content.

Manganese Requirements[6]

Age	RDA for manganese (mg/day)
Children 0–3 years	0.03–1.2
Children 4–8 years	1.5
Children 9–13 years	1.6–1.9
Teenagers and adults	1.8–2.3

REFERENCES

1. Keen, C. L. & Zidenberg-Cherr, S. In E. E. Ziegler & L.J. Filer (Eds.) *Present Knowledge in Nutrition*, 7th Edition. Washington DC: ILSI Press, 1996: 334-343.

2. Linus Pauling Institute. http://lpi.orst.edu/index.html

3. Finley, J. W. & Davis, C. D. Manganese deficiency and toxicity: are high and low dietary amounts of manganese cause for concern? *Biofactors*. 1999;10(1):158-167.

4. Campbell, J. D. Lifestyle, minerals and health. *Medical Hypotheses*. 2001;57(5): 521-531.

5. Johnson, S. The possible crucial role of iron accumulation combined with low tryptophan, zinc and manganese in carcinogenesis. *Medical Hypotheses*. 2001: 57(5): 539-543.

6. Institute of Medicine. Food and Nutrition Board. *Dietary Reference Intakes for Vitamin A, Vitamin K, Arsenic, Boron, Chromium, Copper, Iodine, Iron, Manganese, Molybdenum, Nickel, Silicon, Vanadium, and Zinc*. 2001.

PHOSPHORUS

WHAT IS PHOSPHORUS?

Phosphorus is an essential mineral that is well known for its important role in bones and teeth; however, phosphorous is also integral to the normal functioning of every other cell within the body. It makes up 0.8-1.1% of total body weight, making it the second most abundant mineral in the body. Phosphorous is involved as major structural components in cell membranes such as phospholipids, phosphoproteins and nucleic acids, involved in energy production, components of enzymes, hormones and many other molecules in the body[1,2].

HOW MUCH PHOSPHORUS DO WE NEED?

RDA for phosphorus[3]

Age and gender	RDA (mg/day)
Children 0–6 months	100 (adequate intake)
Children 6–12 months	275 (adequate intake)
Children 1–8 years	460–500
Males 9–18 years	1250
Females 9–18 years	1250
Males 19 and older	700
Females 19 and older	700
Pregnant women	700–1250
Breastfeeding women	700–1250

SOURCES OF PHOSPHOROUS

Almost all foods contain phosphorous. However dairy products, peas, meat, egg, and some cereals and breads are the best sources[3]. In addition, phosphorous, in the form of **phosphoric acid**, is present in many food additives and in many soft drinks [2].

PHOSPHATE DEFICIENCY

Phosphorous deficiencies are rare. Alcoholics may develop hypophosphatemia. Symptoms of **hypophosphatemia** (phosphate deficiency) are bone pain, muscle weakness and bone diseases.

Food sources of phosphorus[2]

Food	Amount of phosphorus (mg)
Skim milk, 8 oz.	247
Plain nonfat yogurt, 8 oz.	383
Mozzarella cheese, 1 oz.	131
Egg, 1 large	104
Cooked beef, 3 oz.	173

Food	Amount of phosphorus (mg)
Cooked chicken, 3 oz.	155
Cooked turkey, 3 oz.	173
Halibut, 3 oz.	242
Salmon, 3 oz.	252
Whole wheat bread, 1 slice	64
Enriched white bread, 1 slice	24
Carbonated cola drink, 12 oz.	44
Almonds, 1 oz.	139
Peanuts, 1 oz.	101
Cooked lentils, ½ c.	356

REFERENCES

1. Arnaud, C. D. & Sanchez, S. D. Calcium and Phosphorous. In E. E. Ziegler & L.J. Filer (Eds.) *Present Knowledge in Nutrition*, 7th Edition. Washington DC: ILSI Press, 1996: pp. 245-255.

2. Linus Pauling Institute. http://lpi.orst.edu/index.html

3. Institute of Medicine, Food and Nutrition Board. *Dietary Reference Intakes: Calcium, Phosphorus, Magnesium, Vitamin D, and Fluoride.* Washington, DC: National Academy Press, 1997: pp. 146-189.

BIBLIOGRAPHY

1. Linus Pauling Institute. Oregon State University. http://lpi.orst.edu/

2. Ziegler, E. & Filer, L. J. (Eds.) *Present Knowledge in Nutrition*, 7th edition. Washington DC: ILSI Press, 1996.

POTASSIUM

WHAT IS POTASSIUM?

Potassium is found almost everywhere in nature, from soil to seawater. This is not surprising, since potassium is an essential mineral needed by all plants and animals on earth. In the human body, the majority of potassium is in the major organs and tissues. It is the third most abundant mineral in the body.

Potassium is extremely important to proper nerve impulse transmission, muscular contractions and maintenance of proper heart rhythm[1,2]. Studies have established that potassium plays a role in preventing hypertension.

POTASSIUM REQUIREMENTS

For the majority of people, a normal diet containing a variety of different foods should provide enough potassium. It is suggested that an adult consume at least 2,000 mg per day. Amounts greater than 3,500 mg per day may have a beneficial effect on hypertension.

POTASSIUM DEFICIENCY

A person suffering from potassium deficiency may experience weakness and fatigue, poor appetite, nausea, slow reflexes, skin problems, mood changes, heart arrhythmias, and high blood pressure.

Several factors may contribute to potassium deficiency. People with high fever, or those who experience fluid loss, stress, shock or trauma, may experience potassium deficiency. Certain medications used in the treatment of high blood pressure may also affect potassium status. Kidney problems, or the overuse of laxatives and diuretics, may also have an impact on potassium levels.

Cohen et al.[3] established an association between mild potassium deficiency and an increased incidence of cardiovascular events in hypertensive patients treated with diuretics. However, the researchers did not establish whether potassium supplementation would decrease the incidence of cardiovascular incidents.

SOURCES OF POTASSIUM

Potassium is abundant in many foods. It is present in fruits, vegetables, whole grains, molasses, fish, and unprocessed meats. Other good sources of potassium are potatoes, spinach, squash, bananas, orange juice and milk.

REFERENCES

1. Linus Pauling Institute. http://lpi.orst.edu/index.html

2. Luft, F. C. Potassium and Its Regulation. In E. E. Ziegler & L.J. Filer (Eds.) *Present Knowledge in Nutrition*, 7th Edition. Washington DC: ILSI Press, 1996: pp. 272-276.

3. Cohen, H. W., Madhavan, S. & Alderman, M. H. High and low serum potassium associated with cardiovascular events in diuretic treated patients. *J Hypertension*. 2001; 19:1315-1323.

BIBLIOGRAPHY

1. Ziegler, E. & Filer, L. J. (Eds.) *Present Knowledge in Nutrition*, 7th edition. Washington DC: ILSI Press, 1996.

SELENIUM

WHAT IS SELENIUM?

Selenium, an essential trace element, is found in animal tissues, principally in two forms: **selenomethionine** and **selenocysteine**.

- Selenomethionine cannot be synthesized by the body, and must be obtained from the diet; it is used by the body in times of selenium shortage.
- Selenocysteine is the biologically active form of selenium[1].

Selenium is a chief component of the body's antioxidant defense systems. An important antioxidant enzyme, **glutathione peroxidase**, contains selenium within its structure. Glutathione peroxidase scavenges for reactive oxygen species that are damaging to the tissue and converts them into harmless molecules of water.

In addition to its role in glutathione peroxidase, selenium is involved in other components of the immune function and in the proper functioning of the thyroid gland[3,4,5]. Selenium was first found to prevent liver necrosis in **Vitamin E** deficient rats.

Recently, much attention has been focused on selenium's anti-cancer properties. Researchers have found that selenium possesses strong antiproliferative properties associated with prostate cancer in humans[6]. However, there is some controversy regarding selenium's role in cancer progression and prevention; further research is needed to establish what role selenium actually plays in protection against cancer.

SELENIUM REQUIREMENTS

Selenium Requirements[9]

Age and gender	RDA (µg/day)
Infants	15–20 (Adequate intake)
Children	20–30
Adults	40–50
Pregnant women	60
Breastfeeding women	70

SELENIUM DEFICIENCY

The diversity of the North American food supply provides adequate dietary selenium levels for most people, and selenium deficiencies are therefore uncommon. However, deficiency can occur in people who are on total **parenteral nutrition** or who have severe problems with gastrointestinal absorption.

Selenium deficiency causes heart problems due to the depletion of selenium-associated enzymes that protect cell membranes from oxidative damage[8]. Deficiency can lead to the development of Keshan disease, which causes malfunctioning and enlargement of the heart. Keshan disease commonly occurs in women and children in certain areas of China[1].

SELENIUM TOXICITY

Selenium can be toxic. However, toxicity occurs only at much greater levels than are normally included in the diet. These overly high levels can result from taking dietary supplements.

Symptoms of selenium toxicity include hair and nail brittleness and loss, gastrointestinal disturbances, skin lesions, tooth decay, a garlic breath odor, fatigue, irritability, and nervous system abnormalities[1,2].

SOURCES OF SELENIUM

The richest sources of selenium are of animal origin. Organ meats and seafood are particularly good sources. Selenium is also found in foods of plant origin. Plants grown in selenium-rich soil are better sources of selenium than plants grown in poor soil.

REFERENCES

1. Linus Pauling Institute. http://lpi.orst.edu/index.html

2. Levander, O. A. & Burk, R. F. Selenium. In E. E. Ziegler & L.J. Filer (Eds.) *Present Knowledge in Nutrition*, 7th Edition. Washington DC: ILSI Press, 1996: pp321-328.

3. Levander, O. A. Nutrition and newly emerging viral diseases: An overview. *J Nutr* 1997;127:948S-950S.

4. Arthur, J. R. The role of selenium in thyroid hormone metabolism. *Can J Physiol Pharmacol* 1991;69:1648-52.

5. Corvilain, B., Contempre, B., Longombe, A. O., Goyens, P., Gervy-Decoster, C., Lamy, F., Vanderpas, J. B. & Dumont, J. E. Selenium and the thyroid: How the relationship was established. *Am J Clin Nutr* 1993;57 (2 Suppl):244S-248S.

6. Venkateswaran, V., Klotz, L. & Fleshner, F. E. Selenium modulation of cell proliferation and cell cycle biomarkers in human prostate carcinoma cell lines. *Cancer Research.* 2002; 62:2540-2545.

7. Early, D. S., Hill, K., Burk, R. & Palmer, P. Selenoprotein levels in patients with colorectal adenomas and cancer. *The Am J Gastroenterol.* 2002; 97(3): 745-748.

8. Burke, M. P. &Opeskin, K. Fulminant heart failure due to selenium deficiency cardiomyopathy (Keshan Disease). *Med, Sci & Law.* 2002;42(1): 10-13.

9. Institute of Medicine, Food and Nutrition Board. *Dietary Reference Intakes: vitamin C, vitamin E, selenium and carotenoids.* National Academy Press, Washington, D.C. 2000.

ZINC

WHAT IS ZINC?

The clinical significance of zinc was first noticed in 1961, when a zinc deficiency resulted in adolescent nutritional dwarfism. This was due to low zinc bioavailability due to the presence of the **antinutrient** phytic acid[1].

Zinc's known importance in human nutrition has expanded dramatically over subsequent years. It plays an essential role in over 100 different enzymes. Researchers have examined the role of zinc in cognition and central nervous system activity, immune development and maintenance, the role of zinc in cancer cell death and the role of zinc in growth development and aging[1,2,3,4]. Zinc also helps in maintaining a normal sense of taste and smell[5].

ZINC REQUIREMENTS

Zinc Requirements[5]

Age and gender	RDA (mg/day)
Infants 0–1 year	2–3
Children 1–8	3–5
Males 9–13	8
Males 14–70	11
Females 9–70	8
Pregnant females	11–12
Breastfeeding females	12–13

ZINC DEFICIENCY

Zinc deficiencies are relatively uncommon. However, there have been cases (especially in children in developing countries) of people born with impaired uptake and transport of zinc.

ZINC TOXICITY

Zinc toxicity can result in gastrointestinal problems, dizziness and nausea[2].

SOURCES OF ZINC

Zinc is present in a variety of foods of plant and animal origin. Foods of animal origin, such as red meat, white meat, and shellfish, are excellent sources of zinc. Plant products such as legumes and nuts are also good sources of zinc. However many plant products high in zinc also contain the antinutrient **phytic acid**, which decreases zinc's bioavailability and renders it unusable by the body.

Certain processing techniques can reduce the level of phytic acid in food products. For example, leavened breads have lower levels of phytic acid because yeast degrades phytic acid[1].

REFERENCES

1. Linus Pauling Institute. http://lpi.orst.edu/index.html

2. Cousins, R. J. Zinc. In E. E. Ziegler & L.J. Filer (Eds.) *Present Knowledge in Nutrition*, 7th Edition. Washington DC: ILSI Press, 1996: pp. 293-306.

3. Prasad, A. S. Zinc: An overview. *Nutrition*. 1995;11:93-99.

4. Heyneman, C. A. Zinc deficiency and taste disorders. *Ann Pharmacother.* 1996;30:186-187

5. Institute of Medicine. Food and Nutrition Board. *Dietary Reference Intakes for Vitamin A, Vitamin K, Arsenic, Boron, Chromium, Copper, Iodine, Iron, Manganese, Molybdenum, Nickel, Silicon, Vanadium, and Zinc.* National Academy Press. Washington, DC, 2001.

WEIGHT-RELATED ILLNESSES

DIABETES

Many people die of diabetes every year. The disease affects 6.2% of the US population[1] and 4.9% to 5.8% of the Canadian population[2]. This translates to 17 million people—11 million in the US and nearly 1.5 million in Canada. In 1999, 450,000 deaths occurred among individuals with diabetes in the US.

Overall, people with diabetes face twice the risk of dying as people without the disease[1]. To make the matter even more serious, many co-morbidities are associated with diabetes. Some complications of diabetes include heart disease, stroke, hypertension, blindness, kidney disease, nervous system damage, amputations and dental disease.

According to the National Institute of Diabetes and Digestive and Kidney Disease, heart disease is one of the leading causes of diabetes related death. People with diabetes are two to four times more likely to have heart disease or have a stroke.

Diabetes is also the leading cause of blindness and renal disease.

People who suffer from diabetes often have mild to severe damage to their nervous system, resulting in impaired sensation or pain in the feet and hands, slow digestion of food, and other nerve problems.

Due to impaired circulation, individuals with diabetes also often undergo amputations of the lower extremeties[1].

WHAT IS DIABETES AND HOW IS IT TREATED?

Insulin-dependent diabetes is associated with the body's inability to make and secrete the hormone **insulin**. Insulin is a key hormone produced by the pancreas.

During the digestion process, carbohydrates are broken down into glucose and other sugars which are the converted to glucose. The body's cells need insulin to convert glucose to energy. When the body is unable to make insulin, glucose builds up in the blood and is excreted in the urine.

There are three main forms of diabetes:

- Type 1 diabetes,
- Type 2 diabetes, and
- Gestational diabetes.

Type 1 diabetes was formerly referred to as **insulin dependent diabetes mellitus** or **juvenile onset**. It occurs as a result of the destruction of the insulin-producing beta cells by the body's immune system. People who have this form of diabetes produce little or no insulin. Treatment involves daily insulin injections and changes to the diet.

Type 2 diabetes is also known as **non insulin dependent diabetes mellitus**. This condition usually starts with **insulin resistance**, a state in which cells cannot effectively use insulin. This results in the pancreas secreting large amounts of insulin. Eventually the pancreas loses its ability to produce insulin. Type 2 diabetes is often associated with obesity and inactivity. Treatment usually involves dietary modification, the introduction of physical activity, and pharmacotherapy. In general, as the body begins to lose weight, it becomes more effective at using insulin. In other words, type 2 diabetes can often be controlled through weight loss.

The third form of diabetes is **gestational diabetes**, which occurs during pregnancy. This form of diabetes is a combination of types 1 and 2, and involves both insulin resistance and an inability to produce insulin. Gestational diabetes is controlled by dietary modification to control blood glucose and by insulin injection. This form of diabetes usually resolves itself after the term of the pregnancy.

REFERENCES

1. NIDDK. National Diabetes Statistics. *General information and national estimates on diabetes in the United States.* 2000. National Diabetes Clearinghouse. http://www.niddk.nih.gov/health/diabetes/pubs/dmstats/dmstats.htm#7

2. Health Canada. *Diabetes in Canada.* Health Protection Branch. http://www.hc-sc.gc.ca/hpb/lcdc/publicat/diabet99/d04_e.html

SLEEP APNEA

Sleep apnea is a sleep disorder characterized by periods without attempts to breathe. This can happen up to 400 times a night[2]. The afflicted person is momentarily unable to move respiratory muscles or maintain airflow through the nose or mouth.

The Greek word apnea means "want of breath." In sleep apnea, the throat and tongue muscles tend to relax during sleep, blocking the passage of air through the throat. When this happens, the brain sends a signal to the breathing muscles, telling them to breathe—often leading the person to wakeup with a snort or a gasp[1].

FORMS OF SLEEP APNEA[1]

There are two types of sleep apnea:

- **Central** sleep apnea occurs when the brain does not send the signals needed for the breathing muscles to initiate breathing.
- **Obstructive** sleep apnea occurs when there is no airflow in or out of the nose, even though efforts are made to breathe.

SYMPTOMS

People suffering from sleep apnea often feel very sleepy during the day, and may fall asleep at inappropriate or dangerous times—for instance, while on the phone or while driving.

WHO IS AT RISK?

People at high risk of developing sleep apnea are those who are overweight, snore loudly, have high blood pressure, or have an abnormality in the upper airway[1].

TREATMENT

Medications are usually ineffective in the treatment of sleep apnea. **Oxygen** and **Continuous Positive Airway pressure (CPAP)** are the most effective treatments. In the case of CPAP, an air blower forces air through the nose to prevent breathing failure. **Dental appliances** can also be used to move the lower jaw for patients with mild to moderate sleep apnea[1].

In more severe cases, **surgery** can be helpful. Although the procedure is not risk-free, surgery can increase the size of the airway and make breathing easier.

Weight loss is a therapy for overweight persons, as it decreases the thickness of the airway walls.

Alcohol and sleeping pills should be avoided by people suffering from sleep apnea because they may aggravate the symptoms[2].

REFERENCES

1. National Sleep Foundation http://www.sleepfoundation.org/publications/sleepap.html#1

2. Familydoctor.org - Health information for the whole family from the American Academy of Family Physicians http://familydoctor.org/handouts/212.html

GOUT

Gout, also known as the "disease of kings," is a disease associated with the inability to metabolize **uric acid**[1]. This results in increased production of, or interferes with the excretion of, uric acid.

Excess uric acid is converted to needle-like **sodium urate** crystals that are deposited in joints and other tissues. The affected joints swell, and may become red and shiny[1].

Gout may be promoted by **obesity**, high intake of alcohol or foods containing purines (such as kidneys, liver, red meat, shellfish, and lentils), and high blood pressure drugs.

One of the most efficient treatments for gout is the anticoagulant **Colchicine**. Unfortunately, the side effects of using Colchicine are quite unpleasant. They include nausea, vomiting and diarrhea. **Non-steroidal anti-inflammatory drugs** (NSAIDs) are also used in the treatment of gout. In addition, dietary restriction of organ meats is recommended in the treatment of gout[1,2].

REFERENCES

1. American College of Rheumatology http://www.rheumatology.org/patients/factsheet/gout.html

2. New Zealand Rheumatology Association
http://www.rheumatology.org.nz/nz08003.htm

THE METABOLIC SYNDROME

WHAT IS METABOLIC SYNDROME?

Insulin resistance, obesity, dyslipidemias and hypertension in combination are referred to as Metabolic Syndrome X. Metabolic Syndrome is a phenomenon that is rapidly developing within North America, although it remains largely undetected. As many as 30% of the adult population is believed to be affected.

Metabolic Syndrome X or Syndrome X (although Syndrome X also refers to a heart condition) refers to a group of health problems including resistance to the action insulin, abnormal blood fat levels, high blood pressure, and often excess weight. All of these conditions significantly increase the risk of heart disease and diabetes.

The following health problems are associated with Metabolic Syndrome X:

- **Central obesity** (excessive fatty tissue in the abdominal region)
- **Glucose intolerance** (insulin resistance)
- **Dyslipidemia** (high serum triglycerides and low HDL cholesterol)
- **Hypertension**

WHAT CAUSES METABOLIC SYNDROME?

It has been recognized that the underlying cause of Metabolic syndrome is insulin resistance. Insulin resistance is a state in which the body becomes less sensitive, or less able to effectively use, the insulin that the body produces.

With a typical North American diet high in carbohydrates, the body is forced to increase production of insulin to compensate for the excess carbohydrate load. The high carbohydrate diet, particularly where it is composed primarily of sugars and lacking in more complex, high fiber carbohydrates, is thought to be a significant contributor to the development of **obesity** and overeating.

Individuals who show insulin resistance for many years are highly susceptible to type 2 diabetes. In addition to diabetes, insulin resistance causes **sodium retention**. Uncontrolled, sodium retention leads to the development of hypertension, another major risk factor for cardiovascular disease. Insulin resistance and diabetes both contribute to abnormal production of lipids by the liver, contributing to the formation of dyslipidemias[1].

WHO IS AT RISK FOR METABOLIC SYNDROME?

If an individual suffers from at least three of the following five medical conditions, they are at risk of developing—or may already have—Metabolic Syndrome X. The family physician should be consulted for proper diagnosis and development of a treatment regime.

- Abdominal Obesity[1]
- High blood triglycerides[1]
- Reduced HDL cholesterol[1]
- High Blood pressure[1]
- Insulin Resistance[1] .

REFERENCES

1. Koenig, V. *Metabolic Syndrome X.* www.stoneyfield.com/HealthyPeople/MetabolicSyndromeX7_02.shtml

BIBLIOGRAPHY

1. American Heart Association. *Syndrome X or Metabolic Syndrome.* 2002. www.americanheart.org

2. Challem, J., Berkenson, B. & Smith, M. D. The Complete Nutritional Program to Prevent and Reverse Insulin Resistance. 2000. John Wiley & Sons, Inc. www.syndrome-x.com

CORONARY HEART DISEASE AND CHOLESTEROL

The incidence of coronary heart disease (CHD) caused by high cholesterol is very high, and is associated with the North American diet. High cholesterol is a function of a fatty diet combined with a sedentary lifestyle.

Unfortunately, unless North Americans make lifestyle changes, CHD and high cholesterol will continue to affect both productivity and lifespan. North Americans must adopt healthful lifestyle habits to lower the risk of CHD and lower cholesterol.

The following table contains a guide to lifestyle changes for a healthy heart presented by the *National Cholesterol Education Program: Adult Treatment Panel III Report.*

HEALTHY LIFESTYLE CHANGES

Summary of Possible Lifestyle Changes[1]

Food items to choose more often
Breads and cereals More than 6 g/day, adjusted to caloric needs Breads, cereals, especially whole grain; pasta; rice; potatoes; dry beans and peas; low-fat crackers and cookies
Vegetables 3–5 g/day fresh, frozen, or canned, without added fat, sauce or salt
Fruits 2–4 g/day fresh, frozen, canned, dried
Dairy products 2–3 g/day Fat free, ½%, 1%, buttermilk, yogurt, cottage cheese; fat free and low-fat cheese
Eggs Less than 2 egg yolks per week; Eat egg whites or egg substitute
Meat, poultry, fish Less than 5 oz. per day Lean cuts of loin, leg, round; extra-lean hamburger; cold cuts of lean meat or soy protein; skinless poultry; fish
Fats and oils Adjust amount to caloric level: Unsaturated oils; soft or liquid margarines and vegetable oil spreads, salad dressings, seeds, nuts
Therapeutic lifestyle change options Margarines containing stanol/sterol; viscous fiber food sources: barley, oats, psyllium, apples, bananas, berries, citrus fruits, nectarines, peaches, pears, plums, prunes, broccoli, Brussels sprouts, carrots, dry beans, peas, soy products (tofu miso)

Food items to choose less often
Breads and cereals Bakery products such as doughnuts, biscuits, butter rolls, muffins, croissants, sweet rolls, Danish pastries, cakes, pies, coffee cakes, cookies Many grain-based snacks, including chips, cheese puffs, snack mix, regular crackers, buttered popcorn
Fats and oils Butter, shortening, stick margarine, chocolate, coconut
Vegetables Vegetables fried or prepared with butter, cheese, or cream sauce
Fruits Fruit fried or served with butter or cream
Dairy products Whole milk, 2% milk, whole milk yogurt, ice cream, cream, cheese
Eggs Egg yolks, whole eggs
Meat, poultry, fish Higher fat meat cuts: ribs, t-bone steak, regular hamburger, bacon, sausage; cold cuts: salami, bologna, hot dogs; organ meats: liver, brains, sweetbreads; poultry with skin; fried meat; fried poultry; fried fish

Recommendations for weight reduction
Weigh regularly Record weight, BMI & waist circumference
Lose weight gradually Goal: Lose 10% of body weight in 6 months. Lose ½ to 1 lb. per week
Develop healthy eating patterns Choose healthy foods (see column 1)
Reduce intake of foods in column 2
Limit number of eating occasions

Recommendations for weight reduction
Select sensible portion sizes
Avoid second helpings
Identify and reduce hidden fats by reading food labels. Choose products lower in saturated fat and calories. Ask about ingredients in ready-to-eat foods
Identify and reduce sources of excess carbohydrates, such as fat free and regular crackers; cookies and other desserts; snacks; and sugar-containing beverages

Recommendations for increased physical activity

Make physical activity part of daily routines

Reduce sedentary time:

Walk, wheel, or bike-ride more, drive less. Take the stairs instead of the elevator. Get off the bus a few stops early and walk the remaining distance. Mow the lawn with a push mower. Rake leaves. Garden. Push a stroller. Clean the house. Do exercises or pedal a stationary bike while watching television. Play actively with children. Take a brisk 10 minute walk or wheel before work, during your work break and after dinner.

Make physical activity a part of recreational activities:

Walk, wheel or jog. Bicycle or use an arm-pedal bicycle. Swim or do water aerobics. Play basketball. Join a sports team. Play wheelchair sports. Golf (pull cart or carry clubs). Canoe. Cross-country ski. Dance. Take part in an exercise program at work, home, school or gym.

REFERENCES

1. National Cholesterol Education Program. *Third Report of the Expert Panel on Detection, Evaluation and Treatment of Cholesterol in Adults (Adult Treatment Panel III Report)*. National Institutes of Health.

SPECIAL TOPICS

HOMOCYSTEINE AND CHOLESTEROL

WHAT IS HOMOCYSTEINE?

Blood cholesterol has been an important focus of health research and media attention because of its role in causing heart disease. Recently, attention has focused on blood **homocysteine** level as an independent risk factor for developing heart disease.

As with cholesterol, high blood concentrations of homocysteine can damage artery walls and contribute to blockage of blood vessels. Homocysteine's potential to damage arteries has been suggested for at least thirty years.

THE HOMOCYSTEINE THEORY

Homocysteine—a sulfur-containing **amino acid**—is a natural byproduct of metabolism. Homocysteine is one of the products formed when the essential amino acid **methionine** is metabolized.

Homocysteine is further metabolized through various metabolic pathways that are dependent on **Vitamin B12**, **folic acid**, and **Vitamin B6**. Normally the metabolic process is very efficient in preventing the accumulation of homocysteine in the blood. However, the high amount of **animal protein** consumed in the North American diet poses a problem.

Animal proteins—in meats, poultry, and dairy products—may contain up to three times as much methionine as plant protein. The digestion of animal proteins therefore creates much more homocysteine to be metabolized. It is this excess homocysteine that causes tissue damage, especially in children with **homocystinuria** or in the arteries in persons of any age.

Dietary studies support a further correlation between homocysteine levels and micronutrient intake. A study of nearly 1,200 elderly people showed that the lower the plasma levels of **Vitamins B6**, **B12** and **folate**, the higher the blood homocysteine level[1,2]. Although these studies have established a correlation between homocysteine levels and cardiovascular disease, the mechanisms at work must be better understood before a true causal link can be established[3].

Other studies have concluded that a mildly to moderately elevated non-fasting plasma homocysteine level is a substantial risk marker for death. The association is strong in deaths from cardiovascular disease, stroke, and other deaths (excluding those caused by cancer)[4,5].

WHAT HOMOCYSTEINE DOES TO ARTERIES

Experiments on arterial segments and on live animals, as well as observations on people with elevated homocysteine levels, have shown that homocysteine can contribute to several processes that lead to **atherosclerosis**:

- Homocysteine generates **superoxide** and **hydrogen peroxide**, both of which have been linked to damage of the endothelial lining of arterial vessels.
- Homocysteine changes coagulation factor level, encouraging blood clot formation. It prevents small arteries from dilating, making them more vulnerable to obstruction by clot or plaque[6].
- Homocysteine causes the smooth muscle cells that support the arterial wall to multiply. This is part of the **atherogenic** process[7].
- In its reactive form, **homocysteine thiolactone**, homocysteine causes platelets to aggregate. This is part of the **clotting** process.

- When infused into baboon arteries, homocysteine causes the endothelial lining cells to actually slough off. This is similar to the behavior of lesions in humans with cardiovascular disease[8].

REFERENCES

1. Selhub, J., Jacques, P. F., Wilson, P. W. F., et al. Vitamin status and intake as primary determinants of homocysteinemia in an elderly population. *JAMA*. 1993;270:2693-2698.

2. Selhub, J., Jacques, P., Rosenber, I. et al. Serum total homocysteine concentrations in the Third National Health and Nutrition Examination Survey (1991-1994): Population reference ranges and contribution of vitamin status to high serum concentrations. *Ann Intern Med*. 1999;131:331-339.

3. Eikelboom, J., Lonn, E., Genest, J. et al. Homocysteine and cardiovascular disease: A critical review of the epidemiologic evidence. *Ann Intern Med*. 1999;131:363-375.

4. Kark, J., Selhub, J., Adler, B. et al. Nonfasting plasma total homocysteine level and morality in middle aged and elderly men and women in Jerusalem. *Ann Intern Med*. 1999;5:321-330.

5. Bostom, A., Rosenberg, I., Silbershatz, H. et al. Nonfasting plasma total homocysteine levels and stroke incidence in elderly persons: The Framingham Study. *Ann Intern Med*. 1999;131:352-355.

6. Wu, K. S., Chook, P., Lolin, Y. I., et al. Hyperhomocyst(e)inemia is a risk factor for arterial endothelial dysfunction in humans. *Circulation*. 1997;96:2542-2544.

7. Tsai, J. C., Perrella, M. A., Yoshizumi, M., et al. Promotion of vascular smooth muscle cell growth by homocysteine: a link to atherosclerosis. *Proc Natl Acad Sci U S A*. 1994;91:6369-6373.

8. Harker, L. A., Slichter, S. J., Scott, C. R., et al. Homocystinuria: vascular injury and arterial thrombosis. *New Engl J Med*. 1974;291:537-543.

SWEETENERS

WHAT ARE SWEETENERS?

Sweeteners are classified as nutritive, or non-nutritive:

- **Nutritive sweeteners**, like table sugar or sucrose (glucose linked to fructose) have calories (4 per gram) and therefore give the body energy.
- **Non-nutritive sweeteners** contain very little to no calories and therefore contribute negligible amounts of energy to the body. These non-nutritive sweeteners are commonly known as sugar substitutes or high intensity sweeteners. These sugar substitutes can be many times sweeter than standard table sugar (sucrose).

WHAT DO SWEETENERS DO?

Nutritive sweeteners contribute a sweet taste, aid in moisture retention, contribute to product tenderness, and contribute to the tactile enjoyment or "mouthfeel" of the product. Non-nutritive sweeteners make food taste sweet, but do not improve the texture or feel of food.

The most common nutritive sweeteners are brown, powdered, granulated, or raw sugars typically derived from sugar canes or sugar beets. Other forms of nutritive sweeteners include corn syrup, honey, the natural sugars found in fruits and juices, maple sugar or syrup, and molasses.

There is a large variety of non-nutritive sweeteners available commercially. These are all dubbed "high intensity sweeteners," as they provide many times the sweetness of an equivalent amount of sucrose. The US Food and Drug Administration (FDA) has approved the following four sugar substitutes for use in foods:

- Saccharin
- Aspartame
- Acesulfame-K, and
- Sucralose.

Saccharin was discovered in 1879 and was originally used to sweeten foods in WWI and WWII. It is a very common tabletop sweetener, served in single serving packages. Due to its many functional attributes, saccharin is used widely to sweeten sodas and baked goods. It can be found under various trade names including Sweet n' Low[2].

Aspartame can be found commercially under such trade names as NutraSweet and Equal. Aspartame is 180 times sweeter than sugar and is commonly used in the sweetening of beverages, breakfast cereals, desserts and chewing gum. Like saccharin, it is available as a tabletop sweetener[2].

Acesulfame-K was approved in 1988 as a tabletop sweetener and is commercially available as Sunett. It is 200 times sweeter than sugar. It is commonly used in baked goods, frozen desserts, candies, and beverages.

Sucralose, also known as Splenda, was approved in 1998 as a tabletop sweetener. It is 600 times sweeter than sugar and is commonly used in baked goods, nonalcoholic beverages, gums, and juices[2].

SAFETY

There has been a great deal of controversy over the safety of artificial or non-nutritive sweeteners and their possible connection to human cancers. **Cyclamate**, an artificial sweetener, received much media attention in the early 1960s because of findings that it caused bladder cancer in laboratory animals, suggesting a possible risk to human safety. These findings prompted the FDA to ban the use of cyclamate in 1969. More recent studies have failed to produce evidence that cyclamate increases cancer risk in laboratory animals. However, other issues have yet to be resolved before cyclamate can be reintroduced in the marketplace.

Early animal studies led to similar conclusions that **saccharin** could cause cancer. However, because researchers conducting large studies involving humans found saccharin to be an unlikely risk factor and concluded that it posed no risk for cancer in humans, the FDA did not ban its use.

Some reports alleged that **aspartame** may be linked to certain brain disorders, including tumors. However, neither studies involving animals nor those involving humans have established a clear link between aspartame intake and the risk of brain tumors[1]. Aspartame is one of the most thoroughly studied artificial sweeteners, and more than 100 toxicological and clinical studies confirm its safety[2].

However, aspartame is contraindicated for individuals with the genetic disorder **phenylketonuria**. Food products containing aspartame must carry a warning declaring that the product contains **phenylalanine**[2].

REFERENCES

1. National Cancer Institute. *Cancer Facts: Artificial Sweeteners.* 1997. Available at: http://cis.nci.nih.gov

2. Henkel, J. Sugar substitutes: Americans opt for sweetness and lite. *FDA consumer.* 1999;33(6):12-16.

FAT SUBSTITUTES

The increased prevalence of obesity in North America has dramatically increased the need for healthier, low-fat products. It is recommended that the diet contain no more than 30% fat and that no more than 10% of this be saturated fat.

Traditionally, bakers have used water, applesauce, and fruit puree as fat replacers to lower the fat content and number of calories in their products. Recent research has found many other forms of fat replacements, from Olean® or olestra to less commonly known non-digestible polysaccharides and products derived from proteins.

Fat substitutes are added to food products with the goal of providing the same flavor and mouthfeel as their higher fat counterparts, but fewer calories. The table below provides a comprehensive list of fat substitutes.

Classification of Fat Substitutes by Nutrient Source, Functional Properties, and Use in Food[1]

Type of fat substitute	Nutrient source (energy density)	Functional properties	Use in food
Derived from carbohydrates			
Polydextrose	Water-soluble polymer of dextrose (1 cal/g)	Bulking and retaining moisture	A wide range of foods, including baked goods, confections, frozen desserts, and salad dressings
Modified food starch	A variety of starch sources (1–4 cal/g)	Modifying texture, gelling, thickening & stabilizing	Processed meats, salad dressing, baked goods, frozen desserts
Dextrin and maltodextrin	A variety of starch sources (4 cal/g)	Modifying texture and bulking	Baked goods, dairy products, salad dressing, sauces, spreads, etc.
Gums and pectin	Zanthan, guar, locust bean, carrageenan, alginates and fruit (virtually non-caloric)	Retaining moisture and modifying mouthfeel	Wide range of products, including baked goods, sauces, and salad dressings
Cellulose	Various plant sources (virtually noncaloric)	Modifying mouthfeel, texture and pouring qualities	Dairy products
Beta Glucan	Soluble fiber extracted from oats (sometimes barley) (1–4 cal/g)	Adding body and texture	Baked goods and a variety of other food products

Type of fat substitute	Nutrient source (energy density)	Functional properties	Use in food
Derived from protein			
Microparticulated protein and modified whey	Denatured or microparticulated protein from egg or milk (1-4 cal/g)	Modifying mouthfeel	Dairy products, spreads, and bakery products
Derived from fat			
Olestra	Sucrose polyester (not absorbed) (noncaloric)	Modifying texture and mouthfeel	FDA approved for use in savory snacks; functionally can be used in place of any fat, including fried foods
Caprenin and salatrim	Carpylic, capric and behenic acid and glycerine or tryglyceride of short and long chain fatty acids (5 cal/g)	Simulating properties of cocoa butter	Confections, baked goods, and dairy products
Mono or diglycerides	Derived from vegetable oil and emulsified with water (9 cal/g; reduces quantity of fat needed)	Adding moisture and modifying texture and mouthfeel	Baked goods, vegetable dairy replacer

RECOMMENDATIONS

The use of fat substitutes is generally regarded as safe. However fat substitutes derived from fat, in particular olestra, can affect fat-soluble vitamin absorption. Fat soluble vitamins are added to products made with olestra to offset this potential effect.

REFERENCES

1. Wylie-Rosett, J. Fat substitutes and Health: An advisory from the nutrition committee of the American Heart Association. *Circulation*. 2002;105;2800-2804.

Appendix B

Colorado on the Move™

APPENDIX B. WHAT IS COLORADO ON THE MOVE?

Colorado is noted for its active lifestyle. It's the only state in the US with an obesity rate of under 15%. The activities that keep Colorado fit and "On the Move" include world-class skiing, mountain camping, and hiking along its extensive mountain and urban trail systems.

Recently, Colorado's Governor challenged its citizens to stay in shape. *Colorado On The Move™* is part of our answer to that challenge. It's a simple program designed to be easy for almost anyone to include in their daily routine; the program's website at www.coloradoonthemove.org offers equipment such as step counters to help citizens (and others) easily track their increase in activity.

The material that follows is from the program's website. It is a terrific way to begin keeping active! It's fun and it's easy.

Colorado On The Move™ can help you increase your levels of physical activity and enjoy many benefits of better health—without drastically changing what you do every day. With *Colorado On the Move™*, you don't need to join a gym or start a workout program. Just incorporate a few extra steps into your day and you'll be on the road to a healthier, stronger body and mind.

Almost everyone can benefit from a few extra steps! First find out how many steps you usually take (this will be your baseline number), then add a few more steps each day. You'll look good, you'll feel great, and you'll see that small changes can make a huge difference in your life.

Remember, when it comes to your health, every step counts.

STEPS MAKE MODERATE EXERCISE EASY

If you're like most people, you usually walk because you need to get somewhere—not because it's exercise. You might think of exercise as something that takes time, equipment, and expensive athletic shoes. *Colorado On the Move™* is ready to challenge the way you think about exercise—and about walking.

You probably already take steps every day. Maybe you walk around in a store, walk to your car, or walk to the corner for a newspaper. That's a great start.

You can take steps almost anywhere—you don't need a gym, a track, or a park. And if you add a few steps here and there to your regular routine, you'll be treating your body to some moderate exercise. And your body will thank you for it!

STEPS ARE SIMPLE TO TAKE AND TRACK

It's easy to get started.

You'll be surprised by how many steps you already take in a day—it's probably more than you think! Use a step counter (visit the website at www.coloradoonthemove.org to get one) to keep track of your average daily steps so you can establish your baseline number. Then when you begin adding steps throughout your day, you'll watch your step numbers grow!

Using a step counter is simple. Clip the matchbox-sized counter to your waist in the morning, push a button, and it will count every step you take. Each evening, record the number from the counter's display onto your log sheet (we've included log sheets in this book).

SET YOUR PERSONAL GOAL

Recommended Goal

To set your goal, take your baseline number and add 2,000 steps to it. For example, if your baseline number is 4,500 per day, use 6,500 as your step goal. 2,000 more steps may sound like a lot, but it should only take 15 to 20 minutes over the course of your day. For instance, it could mean an average of just 5 extra minutes of walking every 3 hours, or 1½ extra minutes every hour. Try it and see how it easy it is!

Don't be frustrated if you have trouble reaching 2,000 steps at first. You might prefer to increase your steps by percentage. For instance, if you're already walking 3,000 steps each day, increasing your steps by 20% will mean 3,600 steps. As adding steps becomes easier, just increase your percentage.

Most people find they can reach 10,000 steps a day within a few weeks of starting the program.

Increasing Your Steps

You'll be surprised how easy it is to add steps to your day. Just do what you already do, but add a few steps here and there!

Once You Reach Your Goal

When you reach your goal, reward yourself! It's great to celebrate your accomplishment with step-counting friends. Then, set a new goal. If you added 2000 steps to your daily routine, add another 2000. If you increased your steps by 20%, and you're feeling good, try adding another 20%.

Now that you know how easy it is to add steps, there'll be no stopping you!

STEPS FIT EVERYONE'S LIFESTYLE

From children to seniors, people have found that increasing physical activity through steps is a fun, easy way to take care of their bodies and improve the quality of their lives.

If you can take one step, you can take a few more. If your body is mobile in other ways (e.g., wheelchair), check out the Step Conversion Chart and adapt the *Colorado On the Move™* program to your lifestyle.

STEP CONVERSION CHART

Almost every physical activity you do will count. Your step counter will calculate your steps in most types of physical activity—including walking, running, tennis, soccer, and many others. When you participate in activities like cycling, swimming, or even lifting weights (light weights make a great wheelchair activity), you can still get step credit.

Use the following conversion table to figure out the "steps" you take during some common non-step exercises. If you don't see your favorite exercise here, go online and visit www.coloradoonthemove.org for more information. If you are physically unable to walk, or if you engage in other types of physical activity, give yourself step credit! Measure your alternate activity "steps" by using this conversion chart.

Step conversion chart

Type of activity	Steps / Minute
Cycling	150
Swimming	150
Weight Lifting	100
Downhill Skiing	150
Rollerblading	200

SAMPLE LOG SHEETS

Step Goal for Week 1-4: ______

Day	Baseline steps	Week 1	Week 2	Week 3	Week 4
Sunday					
Monday					
Tuesday					
Wednesday					
Thursday					
Friday					
Saturday					
Weekly Total					
Daily Average					

STEP FACTS

- 1 mile = 2,000–2,500 steps
- 10,000 steps = 4–5 miles
- Nine holes of golf, no cart = 8,000 steps
- One city block = about 200 steps
- 90-minute soccer game = 8,000–10,000 steps
- Most people walk about 1,200 steps in 10 minutes. (Time yourself to find out how far you walk in 10 minutes!)

EASY WAYS TO INCREASE YOUR STEPS

Every step counts towards your good health and happiness. Think steps—anytime, anywhere.

At Home

- Make the after-dinner walk a family tradition.
- Walk your dog, or offer to walk your neighbor's dog.
- Do a fun family challenge to see who can log the most steps in a week.
- Reward your family for meeting step goals with fun activities.
- Take a walk while your kids are playing sports.
- Walk to your neighbor's or friend's house instead of phoning or e-mailing.
- If you make a phone call, walk while you talk.
- Start a walking club with your neighbors or friends.
- Walk to the television to change the channel.
- Turn off the television and do an active family activity.
- Walk around your house during television commercials.
- Get up and move around once every 30 minutes.
- Try to take half of your goal steps by noon.
- Plan walks into your day. For example, walk with a friend at the beginning of the day, and with your family at the end of the day.
- Plan active weekends (longer walks, scenic hikes, playing in the park).
- Take a walk and pick up litter in your neighborhood or in a park.

On the Town

- Park farther away in parking lots.
- Return your grocery cart to the store.
- Avoid elevators and escalators—try the stairs instead.
- Walk, don't drive, for trips of less than one mile.
- Walk around at the airport while waiting for your plane; avoid the people-movers.
- Take several trips when you're unloading groceries from your car.
- Avoid the drive-through. Instead, walk inside.
- Plan active vacations.
- Hike some of Colorado's (or your state's or province's) beautiful trails.

At Work

- Get off the bus earlier and walk a little of the way to work.
- Take several 10-minute walks during the day.
- Choose the farthest entrance to your building, and walk the long way to your office.
- Host "walking" meetings.
- Walk to a restroom, the water fountain, or a copy machine on a different floor.
- Take a longer route to your meeting.
- Walk a few laps around your floor during breaks, or go outside and walk around the block.
- Walk during your lunch break.
- Walk to a colleague's office rather than calling or sending e-mail.
- Take 5-minute walking breaks from your computer.
- Park farther away in the morning and when you go to lunch.
- Take the stairs rather than the elevator or the escalator.
- Start a break-time walking club with your co-workers.
- Walk while using a speaker or cordless phone.
- Get up and move at least once every 30 minutes.
- Start a *Colorado On the Move* ™ worksite program.